Across the Board

Across the Board: The Mathematics of Chessboard Problems

John J. Watkins

PRINCETON UNIVERSITY PRESS
PRINCETON AND OXFORD

Published by Princeton University Press,
41 William Street, Princeton, New Jersey 08540

In the United Kingdom: Princeton University Press,
3 Market Place, Woodstock, Oxfordshire OX20 1SY

Library of Congress Cataloguing-in-Publication Data

Watkins, John J.
Across the board: the mathematics of chessboard problems/John Watkins.
p. cm.
Includes bibliographical references and index.
ISBN 0-691-11503-6 (acid-free paper)
1. Mathematical recreations. 2. Chess. I. Title.

QA95.W35 2004
793.74—dc22 2003062308

British Library Cataloguing-in-Publication Data

A catalogue record for this book is available from the British Library

This book has been composed in Lucida

Typeset by T&T Productions Ltd, London

Illustrations by Arden Rzewnicki

Printed on acid-free paper ∞

www.pupress.princeton.edu

Printed in the United States of America

10 9 8 7 6 5 4 3 2 1

For Laurel

Contents

Preface

> In play there are two pleasures for your choosing—
> The one is winning, and the other losing.
>
> Lord Byron

Games of all kinds were an important part of my childhood. There is a comfortable pleasure, not unlike that of sitting down with a really good book, in retreating from the real world into a simpler realm where the rules are crystal clear and the objectives well defined. I was taught to play chess one summer by my older brother, Reed, when I was very young. Even then, chess had a special, slightly mysterious, attraction that appealed to me and, that summer, I forced all the kids in my neighborhood to join my newly formed chess club. Gradually, though, my enthusiasm for chess waned and by high school my interests had turned elsewhere.

Meanwhile, when I was about 13 or 14, my brother came home from college and showed me a copy of *Scientific American* that contained a column by Martin Gardner called *Mathematical Games.* Reed knew I liked mathematics and thought I would find his monthly column interesting. Little did he know! Just as for many mathematicians of that era, Gardner's articles have been for me a continuous source of inspiration. Just as I can vividly remember details of the chemistry classroom in which I first heard that President Kennedy had been shot, I still remember the exact place where I was sitting when, as a graduate student, I first read Gardner's article on *RSA* codes. One of the things Gardner teaches us is that mathematics is everywhere—even on the chessboard, a topic to which he returned again and again.

Many years later I had turned to more serious mathematics, first to ring theory and then still later to graph theory, and found myself one summer at an international conference on graph theory held every four years in Kalamazoo, Michigan. I had just the previous year introduced a new undergraduate course on *research in graph theory* at Colorado College and I was looking for a new topic to use in the second iteration of that course for the following year. This turned out to be a moment when several strands of my life came together. At this meeting Stephen Hedetniemi of Clemson University gave an invited talk on *Chessboard Problems.* Within the first five minutes of his talk I knew I had found a perfect topic for my undergraduate course. By the end of his talk I realized that the simple black and white checkered object we know as the chessboard contains within its borders an overflowing font of mathematical ideas and problems. I had just discovered what Martin Gardner knew all along: the mathematical chessboard problems he wrote about so often in his column were so much fun precisely because they were ultimately about serious mathematics.

So, my goal in this book is to share that fun with you by showing you as much as I can of the entire range of mathematics that has been discovered over the years living within that humble chessboard. I do not assume any particular level of training or experience—in either mathematics or chess—on your part. I will try hard to tell you everything you need to know! In turn, I hope you will join in the spirit of discovery and try to attempt most, or certainly at least some, of the many problems that I have sprinkled throughout the text, and whose solutions appear at the end of each chapter.

Across the Board

CHAPTER ONE

Introduction

> We are no other than a moving row
> Of Magic Shadow-shapes that come and go
>   Round with the Sun-illumined Lantern held
> In Midnight by the Master of the Show;
>
> But helpless Pieces of the Game He plays
> Upon this Chequer-board of Nights and Days;
>   Hither and thither moves, and checks, and slays,
> And one by one back in the Closet lays.
>
> Omar Khayyám, *The Rubáiyát*

This book is about the chessboard. No, not about *chess*, but about the board itself. The chessboard provides the field of play for any number of games, both ancient and modern: chess and its many variants around the world, checkers or draughts, Go, Snakes and Ladders, and even the word game Scrabble. Boards for these games come in many sizes: 8×8 boards for chess; 8×8 and 10×10 boards for checkers, depending on what part of the world you are in; 10×10 boards for Snakes and Ladders; 15×15 boards for Scrabble; 18×18 boards for Go; and even non-square sizes such as 4×8 and 2×6 boards for Bau and Owari, two games that are widely played in various forms and under several different names in Africa.

In some board games there are no special colors given to individual squares of the board, all squares are the same, but in other games the color of individual squares can be very important. The familiar checkered black and white or black and red alternating coloring of the squares in chess or checkers are the best examples of this. Alfred Butts spent years during the

1930s and 1940s tinkering with the coloring of the board for the game that eventually became Scrabble, deliberating on exactly where the double and triple valued squares should go, what their colors should be, and how many of each type he wanted. We, too, will discover that ingenious colorings of the squares on an otherwise ordinary chessboard can pay surprising dividends.

Board games, like other of our games, are usually in some form a metaphor for life itself. Chess, of course, is about war, conquering your enemy and protecting your king; Go is about the territorial imperative; Snakes and Ladders a morality play for children about how to achieve Nirvana. Omar Khayyám, in the two quatrains I quoted for you from *The Rubáiyát*, saw in such games a reflection of our lives as mere 'pawns' in a game run by the 'Master of the Show', a theme picked up by Shakespeare 300 years later in *As You Like It*, the play with which he possibly opened the Globe Theatre in 1599:

> All the world's a stage,
> And all the men and women merely players;
> They have their exits and their entrances,
> And one man in his time plays many parts,
> His acts being seven ages.

Even now, in an age, and especially in a society, such as ours, in which the problematic notion of free will is simply taken for granted, this is still an idea that carries a chilling weight.

However, this is not a book about such games, and still less is this book concerned with the cultural significance of the games we humans play. Instead, this book is about the game board itself, the simple grid of squares that forms such a common feature of games played around the world, and, more importantly, about the mathematics that arises from such an apparently simple structure. Let us begin with a puzzle.

Guarini's Problem

The earliest chessboard puzzle that I know of dates from 1512, almost 500 years ago. This puzzle is known as *Guarini's Prob-*

Figure 1.1 Guarini's Problem: switch the knights.

lem and involves four knights, two white and two black, at the four corners of a small 3×3 chessboard. The white knights and the black knights wish to exchange places. Their situation is shown in Figure 1.1. A knight can move on a chessboard by going two squares in any horizontal or vertical direction, and then turning either left or right one more square. Since, in this problem, each knight will have only two moves available from any position, this is a very simple puzzle to solve, even by trial and error. Still, it is somewhat harder than it might look at first glance, so I urge you to try to do it for yourself before reading on.

If you solved this problem you undoubtedly observed its underlying basic structure. Note that this rather simple structure comes from two things: the geometry of the board itself along with the particular way in which knights are allowed to move. In Figure 1.2, the structure of the possible knight moves is first exhibited explicitly, on the left, by lines that are drawn to connect any two squares of the board between which a knight can move. This diagram, or *graph*, can then be 'unfolded', as shown on the right in Figure 1.2, and the underlying structure of the graph immediately emerges. The graph, which looked somewhat complicated to us on the left, turns out to consist of only a single cycle, and so the solution to our puzzle is now completely clear. In order to exchange places, the knights have no choice at all. They must march around this cycle, all in the same direction, either clockwise or counterclockwise, until their positions are exactly reversed. This graphical solution, of course, still has to be translated back to the original 3×3 chessboard, but this final step is quite straightforward.

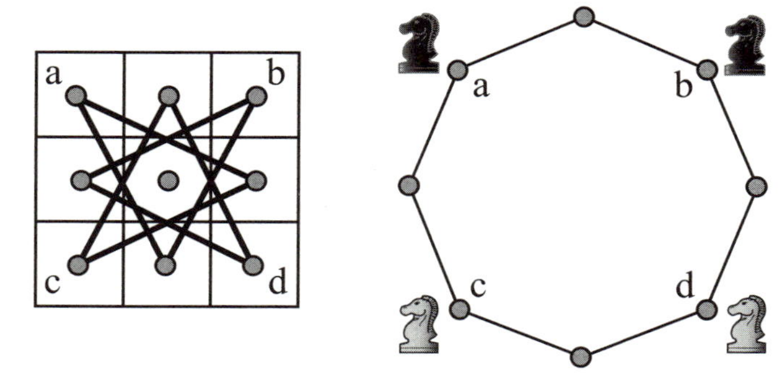

Figure 1.2 Ignoring the chessboard and unfolding the graph.

I like Guarini's Problem in part because it is so very old, but also because it is such a nice illustration of the way in which mathematical abstraction can clear away the messy details of a problem and lead us gently toward a solution. In the case of Guarini's Problem, the first level of abstraction avoids for us the need to keep worrying about the awkward business of how a knight moves on a chessboard by the simple device of drawing lines on the board to represent these moves. This allowed us to forget about chess completely. The next level of abstraction is to forget also about the actual board and focus instead entirely on the diagram. Then, the final level of abstraction is to eliminate the clutter inherent in the diagram simply by unfolding it. This general process of turning a problem into a diagram is so useful and so natural that an entire area of mathematics, now called *graph theory*, has evolved that is dedicated to studying the properties and uses of such diagrams, called *graphs*. This book, then, is really a book about graphs in disguise. Usually, explicit graphs such as the one drawn in Figure 1.2 will be kept offstage during our drama, but rest assured, they are always there and ready to appear at a moment's notice should we ever have need of them.

Problem 1.1 In Figure 1.3 we now have six knights, three white and three black, on opposite sides of a 4×3 chessboard. Find the minimum number of moves required for these knights to exchange places. This variation of Guarini's Problem appeared in *Scientific American* in the December 1979 issue. A solution

Figure 1.3 Switch the knights.

was given the following month. Remember though that you can, if you like, find solutions to these problems at the end of each chapter.

The Knight's Tour Problem

Since we have just studied a problem involving knights, and we now know how they move, let's continue with them for the moment. Note that in Guarini's Problem the center square of the 3×3 chessboard is inaccessible to any of the four knights, but that otherwise a single knight could visit each square of the board exactly once and return to its original position merely by making a complete circuit of the cycle graph in Figure 1.2. What about on a larger board, such as a 4×4 board or an 8×8 board, might a knight be able to do a tour of the entire board, that is, visit every square exactly once and return to the start? The answer is: well, sometimes *yes*, and sometimes *no*.

The general question of which chessboards have a knight's tour is known as *the Knight's Tour Problem*. This famous problem has a long and rich history. The Knight's Tour Problem dates back almost to the very beginnings of chess in the sixth century in India and will in fact be one of the main topics considered in this book.

You might wish to try the following problem before continuing further. The key feature of a knight's tour is that a knight visits each square of the board exactly once. A tour can be *closed*, meaning the knight returns to its original position, or it can be *open*, meaning it finishes on a different square than it started. Unless otherwise indicated, the word 'tour' in this book will always mean a closed tour.

Problem 1.2 The smallest chessboards for which knight's tours are possible are the 5×6 board and the 3×10 board. Find a tour for each of these boards. The smallest board for which an open tour is possible is the 3×4 board. Find an open tour for this board.

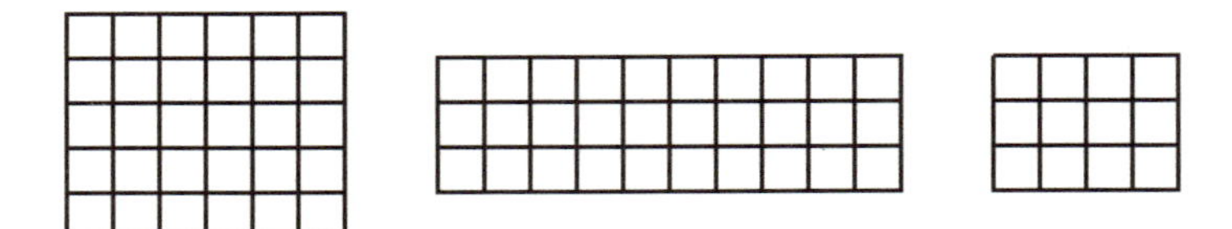

Figure 1.4 Find closed tours for the 5×6 and the 3×10 boards; and an open tour for the 3×4 board.

Leonhard Euler, perhaps the most prolific mathematician of all time, did extensive work on the Knight's Tour Problem. One particularly attractive knight's tour for the 8×8 chessboard, done by Euler in 1759, is shown in Figure 1.5. What is especially interesting about this particular tour is that Euler first does an open tour of the lower half of the board, starting at square 1 and ending at square 32. He then repeats exactly this same tour, in a symmetric fashion, for the upper half of the board, starting at square 33 and ending at square 64. Note that Euler has also very carefully positioned both the beginning and the end of these two open half-tours so that they can be joined together into a tour of the entire chessboard.

This tour of Euler's tells us something very useful about the Knight's Tour Problem, at least we now know for sure that the 8×8 chessboard has a knight's tour. On the other hand, we already know that not *all* chessboards have knight's tours. For example, as we have seen, the 3×3 board can't have a tour for the simple reason that the center square is inaccessible. A more interesting, and much more surprising, example is the 4×4 board. A quick glance at the graph in Figure 1.6 shows

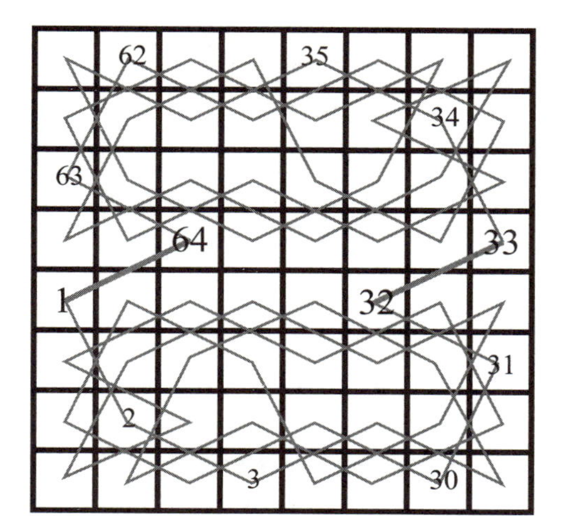

Figure 1.5 Euler's tour of the chessboard.

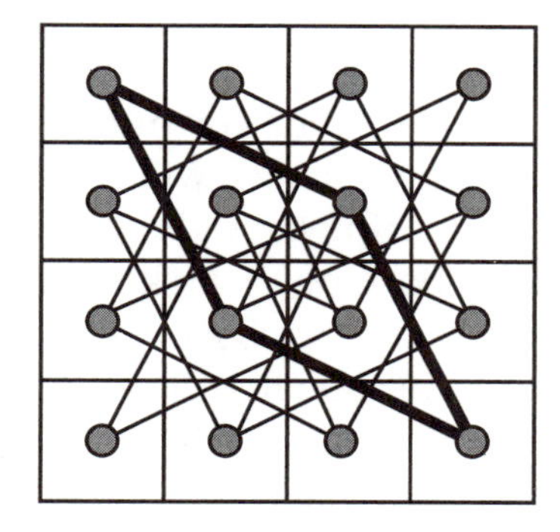

Figure 1.6 Why a 4×4 board has no tour.

why a knight's tour is impossible for this chessboard. The two bold lines, or *edges* as they are usually called in graph theory, coming from the upper left-hand corner represent the only possible way for a knight to get into or away from that particular corner square. Thus, both of these bold edges *must* be a part of any knight's tour of the entire board. For exactly the same reason, the two bold edges coming from the lower right-hand corner must also be part of any tour of the entire board. But, as we can easily see, these four bold edges form a small closed cycle visiting just four squares, which means that they cannot simultaneously also be part of a larger tour of the entire board.

So, because of this unavoidable conflict, a tour of the entire 4×4 board is impossible.

Are there any other chessboards on which a knight's tour is impossible? In order to answer this question, it is going to turn out to be useful for us to color the squares of our chessboards, as we shall see.

Coloring Chessboards

Omar Khayyám used the black and white 'chequer'-board pattern of the chessboard to represent nights and days in the passage I quoted earlier from *The Rubáiyát*, thereby invoking a disturbing sense of the endless cyclic repetitiveness of life. These days, we see the checkerboard pattern used so often for decorative purposes—on taxicabs, on table cloths in Italian restaurants, for tiling floors, for the checkered flags at auto races—that it will seem quite amazing to us how wonderfully useful this pattern also turns out to be for purely mathematical purposes. This is a theme to which we will return again and again. But, for now, let's use coloring to see how to find *infinitely* many chessboards that can't possibly have knight's tours!

Up to this point, of course, we have not bothered to color the squares on any of our chessboards. But now, look at the 5×5 chessboard in Figure 1.7, where we have colored the squares in the usual way. The main thing to note is that a knight on this board always moves from a square of one color to a square of the other color. You should verify this for yourself. Therefore, a knight's tour necessarily alternates between black squares and white squares; and so, a knight's tour must consist of an equal number of black squares and white squares. But, the 5×5 chessboard doesn't have an equal number of black squares and white squares! In fact, our 5×5 chessboard has 12 white squares and 13 black squares. And so a tour of the 5×5 chessboard is impossible.

The same argument, of course, works for any 'odd' chessboard, such as a 13×17 board, where both of the dimensions are odd, simply because in that case the total number of squares would also be odd, and so the number of black squares could

1	14	9	20	3
24	19	2	15	10
13	8	23	4	21
18	25	6	11	16
7	12	17	22	5

Figure 1.7 An open tour of the 5×5 board.

not be equal to the number of white squares. Thus, there are infinitely many chessboards for which a knight's tour is impossible.

Playing on Other Surfaces

It may have occurred to you, however, that just because there isn't a *closed* knight's tour for a 5×5 chessboard doesn't mean that there can't be an *open* knight's tour. After all, couldn't an open tour begin on a black square and then also end on a black square and in that way successfully alternate 13 black squares with 12 white squares? This is indeed possible! One such tour is shown in Figure 1.7.

This particular open tour suggests a natural extension of the Knight's Tour Problem, or at least it suggests something that might seem natural to anyone who has spent a significant amount of time playing video games. What if we change the rules for the knight and now allow the knight to move off the bottom of the board and then reappear at the top, exactly as happens in some video games? Or if we allow the knight to go off of one side of the board and instantly reappear at the other side? This is not an idea that would have seemed particularly natural to Euler in the middle of the eighteenth century when he was playing with knight's tours, but it certainly seems natural to us now. With this new-found, computer-aided freedom, the poor knight who was previously stuck at square 25 in Figure 1.7 can now happily return to square 1 in a single move, thus closing the tour.

In this particular situation, a mathematician might say we were playing chess on a *flat torus*. As we shall see later, this is actually the same as playing chess on the surface of a doughnut! Since I intend for us to be investigating knight's tours on a wide variety of strange surfaces like the torus, you should perhaps warm up a bit first on the following problem.

Problem 1.3 Find a knight's tour on the surface of a $2 \times 2 \times 2$ cube. This needs some explaining. First of all, this cube has only four squares on each face and so if a knight is to move at all it must be able to pass over an edge of the cube onto another face. So, we will agree ahead of time that a knight moves over an edge of the cube without even noticing it. The various possible knight moves will be far easier to visualize with an actual cube in your hand, such as a *Rubik's Mini Cube*, and using a water-based pen to record the moves. A reasonably good alternative is to use the unfolded diagram of the cube shown in Figure 1.8 as a visual aid. For clarity, we should also agree on exactly what constitutes a knight's move. I earlier described this move as being 'two squares in one direction followed by one more square either left or right'. On a flat board this happens to be the same as saying 'one square in one direction followed by two more squares either left or right'. But on a $2 \times 2 \times 2$ cube these two 'moves' don't always turn out to be the same. Since it is pretty hard to argue that either one or the other of these two 'moves' is the *real* knight's move, I will arbitrarily decree that you can use both. You might want to check that a knight at any position, therefore, has a total of ten possible moves. With that much mobility, this shouldn't be a hard problem!

The Domino Puzzle

We have just seen how the standard checkerboard pattern of alternating black and white squares can be used to show that chessboards having odd dimensions cannot possibly have knight's tours. Here, now, is a clever, and quite well known, puz-

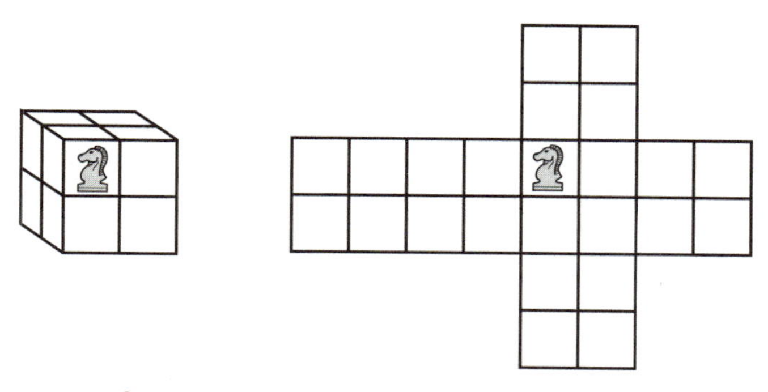

Figure 1.8 An unfolded $2 \times 2 \times 2$ cube.

zle that also illustrates the usefulness of coloring. This puzzle involves dominoes. You may well have seen this puzzle before, but it still provides us with a good introduction to a large class of interesting geometric problems. A *domino*, for us, is just a 1×2 rectangle. We are not interested in any spots it may have, just its one-by-two rectangular shape. If the size of our dominoes exactly matches the size of two adjacent squares on our chessboard, then we can easily see that an 8×8 chessboard can be 'covered' with 32 dominoes. No problem!

Here is the well-known *Domino Puzzle*: remove two diametrically opposite corner squares from an 8×8 chessboard; can you cover the 62 remaining squares of the chessboard with 31 dominoes? It seems as if you should be able to do this, after all, $2 \cdot 31 = 62$, but I do not recommend spending very much time trying to do it—well, maybe a little—because you are doomed to failure.

Instead, try to visualize the checkerboard pattern of the 8×8 chessboard, or, better yet, look at an actual chessboard. All you have to do is note that the two diametrically opposite corner squares that have been removed from the original chessboard have the same color! So, we now have 32 squares of one color, but only 30 squares of the other color. Clearly, then, since a domino covers one white square and one black square, it is impossible to cover the remaining 62 squares with 31 dominoes.

I have always observed that what makes the Domino Puzzle so successful as a puzzle is that people tend to concentrate mostly on the rather bizarre geometric shape of the board once the two opposing corners have been removed, so they approach the problem geometrically. Moreover, the checkerboard pattern is such a familiar design in our daily lives, one which is so frequently encountered that we hardly notice it, and so the color of the squares in this puzzle seems irrelevant.

On the other hand, what if we start over with this same puzzle and remove, say, the upper left-hand corner square and the upper right-hand corner square? Then, as we expected all along, it is a piece of cake to cover the remaining board with 31 dominoes. What if we make things a little harder and instead remove a single black square and a single white square from random positions on the board? Can we still always manage to cover the remaining board with 31 dominoes? Happily, the answer is: *yes*. In other words, the following remarkable theorem is true.

Theorem 1.1 (Gomory's Theorem) *If you remove one black square and one white square from anywhere on the 8×8 chessboard, then you can cover the remaining board with 31 dominoes.*

It is surprisingly easy to give a proof of this theorem. In order to do so, I first need to return to the idea of a tour of a chessboard. This time, however, I want to be thinking of a rook's tour rather than a knight's tour. A rook can move in either a horizontal or a vertical direction and it can also go as many squares as it wants. For our purpose, which, after all, is touring chessboards and visiting squares, it is best to think of the rook as moving just one square at a time. In other words, when a rook moves along a row or column it 'visits' every square along the way. Armed with this knowledge of how a rook moves, I am sure you will have no trouble finding a rook's tour of the 8×8 chessboard. Spend a moment and do this. You probably came up with something like the rook's tour illustrated in Figure 1.9.

Now, using the rook's tour in Figure 1.9, we can give the easy proof of Gomory's Theorem I promised you. First, I suggest you try to ignore completely the fact that this is happening on a chessboard at all and instead just think about the tour itself, which, after all, is just one very long, very big cycle with the rook

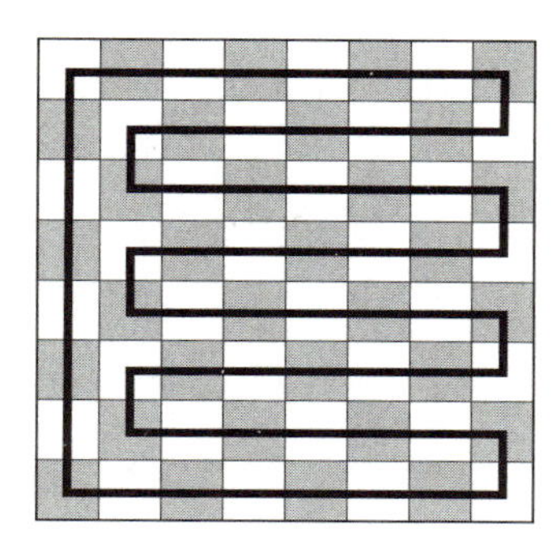

Figure 1.9 A rook's tour of the 8×8 board.

marching merrily along from square to square: *black square, white square, black square, white square, black square, white square,* until finally the rook gets back to the original starting square. Now, if a black square is removed from anywhere along this cycle, then the cycle stays connected in one piece, but is now open with a white square at each end. If a white square is now removed from somewhere along this piece, then we will be left with two pieces (unless, of course, we remove one of the two white ends, but then the argument becomes even easier). We can see that each of these two pieces is now white at one end and black at the other end. So, each individual piece can be covered with a string of dominoes, each domino covering one white square and one black square, and we are done with the proof.

If this argument wasn't clear to you, try thinking of a necklace with 32 black beads alternating with 32 white beads. Cut one black bead out of the necklace and visualize what you are left with, and then also cut one white bead out and visualize the two remaining pieces. That's all there is to it.

There were two keys to the unlocking of Gomory's Theorem: first, our very good friend the standard checkerboard coloring of the chessboard; and, second, the extremely convenient fact that the 8×8 chessboard has a rook's tour. Figure 1.10 presents several attractive rook's tours of the 8×8 board in a slightly different form, including the original tour from Figure 1.9 on the left in this new form. You may want to look for a few more tours. There are lots of them.

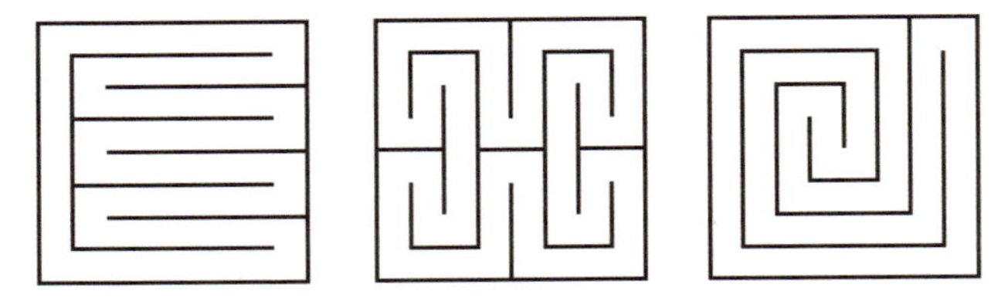

Figure 1.10 Three rook's tours of the 8×8 board.

We will return later in the book to the topic of covering chessboards with 'ominoes' in all manner of sizes: monominoes, dominoes, trominoes, tetrominoes, pentominoes, and so on. In the meantime, here is another problem involving 'ominoes' for you to work on.

Problem 1.4 Figure 1.11 shows the five tetrominoes, that is, the five connected figures that cover four squares of a chessboard. Note that two of these tetrominoes look different if you turn them over, but they are still considered to be the same tetromino. If you take sixteen of the straight 1×4 tetrominoes, you can easily cover the 8×8 chessboard with them. Similarly, for the square 2×2 tetromino, with sixteen of them you can also easily cover the 8×8 board. Which of the other three tetrominoes can be used to cover the 8×8 board in this way, that is, using sixteen copies of the same tetromino? Here is a slightly different covering problem: can you cover the 4×5 board using each of the five tetrominoes? What about a 2×10 board? Don't forget that you can place a tetromino with either side facing up.

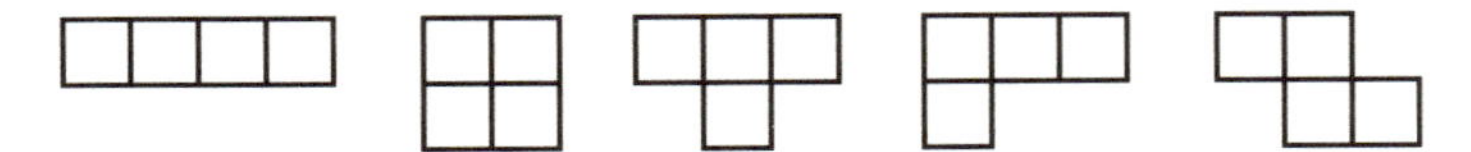

Figure 1.11 The five tetrominoes.

Domination

In the previous section, we talked about a certain kind of 'covering' for chessboards in which a chessboard was literally covered by various geometric shapes such as dominoes or tetrominoes. I'd like to turn now to a consideration of a very different kind of 'covering' for a chessboard. We begin our discussion with

Figure 1.12 Four queens almost cover the entire board.

queens. The queen is by far the most powerful piece in chess, combining as it does the strength of the rook, which moves any distance it likes horizontally or vertically, with that of the bishop, which moves any distance it likes diagonally. Depending upon which square it is placed on, a single queen can *cover* or *control* anywhere from 22 to 28 squares of the chessboard. Now *that* is power.

In fact, the four queens in Figure 1.12 together control almost the entire board, only the two shaded squares escape their control. (I can't resist interrupting myself to point out that if the board in Figure 1.12 were on a flat torus—where, remember, queens could go off the board on one side and reappear at the opposite side—then the two shaded squares would already be covered! Do you see how? In other words, the entire board would be covered by just these four queens. End of interruption, we now return to regular chessboards.)

Since only two squares are uncovered in Figure 1.12, it is easy to place a fifth queen on the board so that the entire board is covered or controlled by these five queens. Unfortunately, it is not at all easy to prove that you actually *need* five queens in order to cover the 8×8 chessboard. This *is* the case, however. We will come back to this and other questions about queens later in the book.

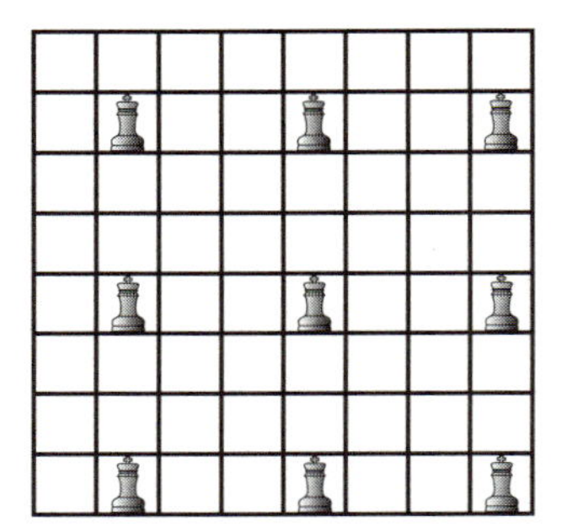

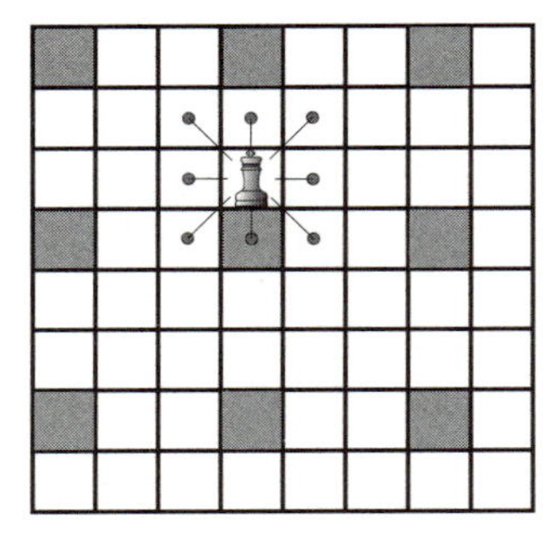

Figure 1.13 Nine kings are required to cover the board.

In the meantime, because it will prove to be a far easier question to answer, let us ask the same question about kings: how many kings do you need to cover an 8×8 chessboard? The king, like the queen, can move in any direction, but is far more limited than the queen, being able to move only one square, that is, to an immediately adjacent square. Thus, a king can cover at most nine squares, namely, the 3×3 portion of the board in which it is centered. It is easy to see, then, that nine kings can be placed so as to cover an 8×8 chessboard, for example, as arranged in the diagram on the left in Figure 1.13.

But, do you *need* nine kings in order to cover the board? The answer is: *yes*, you do. This turns out to be easy to show. Consider the nine shaded squares in the diagram on the right in Figure 1.13. No matter where you might decide to place a king on this board, a single king would cover at most *one* of these shaded squares. In other words, a king could never cover two of the shaded squares at the same time. Therefore, you need at least nine kings merely to get the shaded squares covered, and hence, you need at least nine kings to cover the entire board.

Note that in addressing the question: 'how many kings do you need to cover an 8×8 chessboard?', we had to do two different things in order to conclude that 'nine' is the correct answer. We had to demonstrate that the answer was 'nine or less' by actually covering the board with nine kings. We also had to demonstrate that the answer was 'nine or more' by some other kind of ad hoc argument; in this case, a rather clever argument involving shaded squares did the trick. Since we could do both of these things, the answer ended up being 'nine'.

In the kind of covering problems we have just been looking at we are seeking the *minimum* number of pieces of a given type needed to cover or control the entire board. How many bishops do you need to cover a chessboard? How many knights? These are examples of a large and important category of problems called *domination problems* in graph theory. Domination problems are being energetically investigated by mathematicians all over the world these days, in large part because they arise so naturally in a large number of real-world applications. However, it is worth remembering that domination problems first originated when people began asking 'covering' questions about chess pieces. Even the aggressive sound of the term 'domination' should serve to remind us that domination in graph theory is a topic that had its origins in chess, which, after all, began so many centuries ago as a game of war.

Independence

We now turn to another large and important category of problems. In these problems we will ask questions that are similar to those that we have just met in domination theory, but that have a kind of reverse spin. For example, what is the *maximum* number of knights that you can place on an 8×8 chessboard so that no two knights attack one another? You might want to think about that for a moment before continuing. We will call a group of knights *independent* if none of them attacks any other. So, questions like the one we have just posed are referred to as *independence problems.*

In order to answer this particular question about placing independent knights on a chessboard, we have to do two different things, just as was required of us in answering domination problems. First, we have already noted that a knight on, say, a black square, can only move to a white square, so if we place 32 knights on the 32 black squares, then no two knights can attack one another. So, these 32 knights are independent. It would seem pretty obvious that surely this must be the best that we can do, but how do we go about proving it? That is the second thing we have to do.

Here is one way to do that. Thanks to the work of Euler, we have already seen that the 8×8 chessboard has a knight's tour and we have also observed that this tour alternates between black and white squares. Since we can't place two independent knights on 'adjacent' squares anywhere along this tour, we can place *at most* 32 independent knights along the tour. So, 32 is the best that we can hope to do. This argument gives us a bonus that we weren't even looking for: because the tour alternates between black and white, the 32 knights must either all go on the black squares or they must all go on the white squares. In other words, the easy placement we first thought of—that is, putting all the knights on one color—was, in fact, the only solution.

Here is an alternative argument that 32 is the *best* that we can do. First, divide the board into eight 2×4 rectangles. Since a knight placed anywhere within one of these 2×4 rectangles covers exactly two squares in the 2×4 rectangle, *at most* four independent knights can be placed in each rectangle. But there are eight such rectangles, and so we can't possibly fit more than 32 independent knights on the chessboard.

We will return to both domination problems and independence problems in considerable detail later, but for now, here is a problem for you to do.

Problem 1.5 Place eight bishops on the 8×8 chessboard so that they *dominate* the entire board. Place fourteen bishops on the 8×8 chessboard so that they are *independent*. In each case, this is the best that can be done. A reminder: a bishop moves as many squares as it likes along a diagonal path. Note that a bishop always remains on squares of the same color.

Graeco-Latin Squares

We have at this point introduced the main topics with which we will be concerned in this book: knight's tours, domination, independence, coloring, geometric problems, chessboards on other surfaces, and even polyominoes. But a subject as broad—and as vaguely defined—as 'chessboards' also contains within its scope plenty of room for a large number of miscellaneous topics and problems that, while not particularly easy to categorize, are nonetheless worthy of our attention. I'd like to conclude this introduction by turning briefly to such a topic now.

The very first time I saw one of Martin Gardner's *Mathematical Games* columns was in the November 1959 issue of *Scientific American*. The topic of this column was Graeco-Latin squares. They even had a painting on the cover of that issue that had been inspired by his column, a large 10×10 grid that looked like something that might have been painted by Piet Mondrian.

That's about all I'm going to tell you about Graeco-Latin squares for now. It turns out that the painting on the cover of the magazine, and Gardner's column on the inside, were nothing less than a stunning announcement that a very famous conjecture that had originally been made by Leonhard Euler and that had survived unscathed for nearly 200 years had finally just crashed spectacularly to defeat. As to what a Graeco-Latin square actually is, I'll let you discover that for yourself by doing a puzzle that was popular in the eighteenth century, called *Ozanam's Problem.*

Problem 1.6 Take all four aces, all four kings, all four queens, and all four jacks from a single deck of playing cards. Arrange these sixteen cards in four rows and four columns so that each row and each column contains all four card values as well as all four suits. If you succeed in doing this, you will have created a *Graeco-Latin square of order 4.* Martin Gardner even tossed in an extra challenge: arrange the cards so that the two diagonals of the square also satisfy this same condition.

Solutions to Problems

Solution 1.1 In Figure 1.14 we first draw the knights graph in its natural position on the chessboard and also at the same time number the squares of the board in a convenient fashion. We then, of course, unfold the graph.

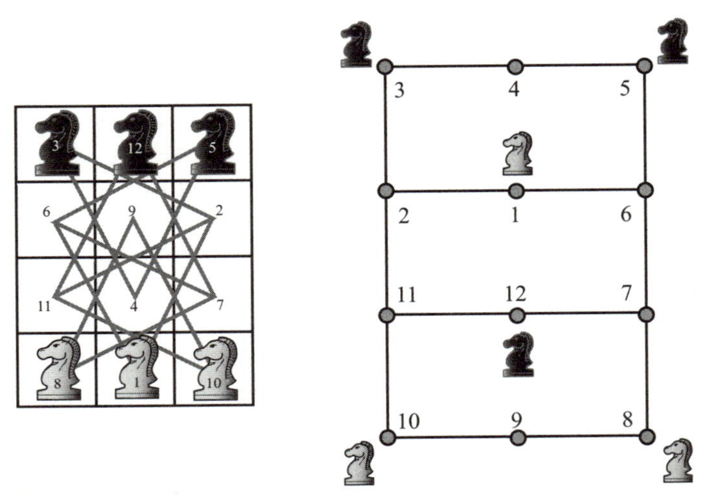

Figure 1.14 Switch the knights in sixteen moves.

We can immediately see that it takes seven moves just to move the three black knights into the correct position; and, by symmetry, there is really only one way to do this: bring the knight at 5 straight down to 8 and move the other two knights two squares each, that is from 3 to 1 and from 12 to 10. And so, we might hope to do the complete switch of all the knights in just fourteen moves, seven for each color, but unfortunately we can't move both the black and the white knights in these seven moves without the knights getting in each other's way. What we can do however is have one knight take a single step backwards at an appropriate time to clear the way for an opposing knight. The result will be a loss of only two moves, which is the best we can hope for.

Here, then, is a sixteen-move solution: the black knight at 5 moves down to 7; this allows the white knight at 1 to move into place at 5, which allows the black knight at 3 to move into place at 1, which then allows the white knight at 10 to move straight into place at 3, and in turn the black knight at 12 to move into place at 10. So far, this has been perfectly efficient, but now the black knight at 7—the polite knight, if you will—steps back to 6, allowing the white knight at 8 to move into place at 12, before continuing his otherwise straight path down to 8.

Solution 1.2 The solutions for this problem are given in Figure 1.15.

5	26	1	16	11	20
2	15	4	19	30	17
25	6	27	12	21	10
14	3	8	23	18	29
7	24	13	28	9	22

26	29	2	21	8	23	6	17	14	11
1	20	27	24	3	18	9	12	5	16
28	25	30	19	22	7	4	15	10	13

1	4	7	10
12	9	2	5
3	6	11	8

Figure 1.15 Tours of the 5×6 and the 3×10 boards, and an open tour of the 3×4 board.

Solution 1.3 The strategy here is to do an open tour of the front three faces (squares 1–12 in Figure 1.16). Then repeat this exact same open tour on the back three hidden faces (squares 13–24). Finally, these two open tours can be linked with knight's moves at squares 12–13 and at squares 24–1 to complete the tour of the entire $2 \times 2 \times 2$ cube. This particular solution of course imitates Euler's tour of the 8×8 board given in Figure 1.5.

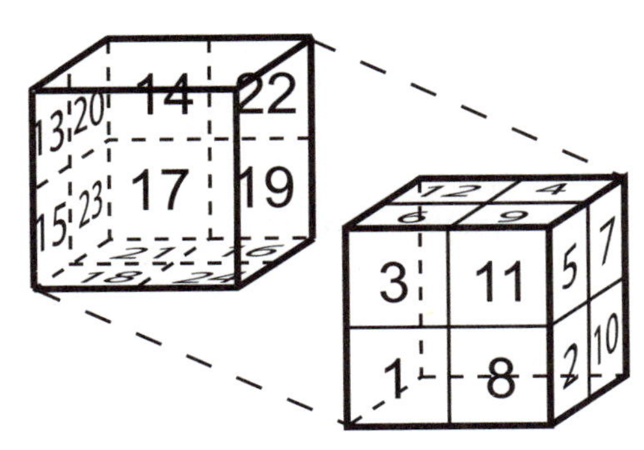

Figure 1.16 The front and back sides of a $2 \times 2 \times 2$ tour.

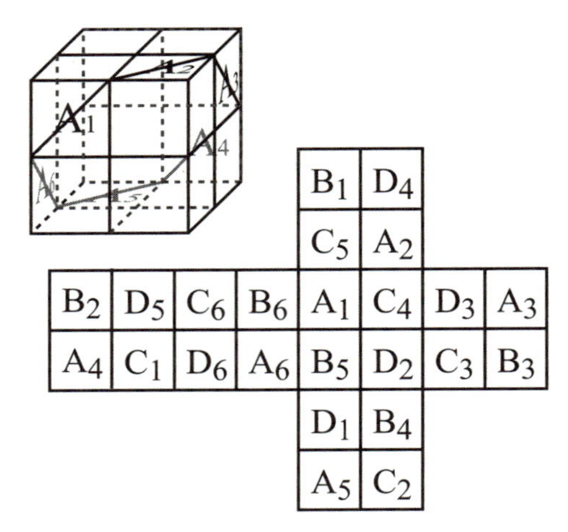

Figure 1.17 A hexagonal cross-section and four mini-tours.

Another solution uses an idea given by Ian Stewart in *Another Fine Math You've Got Me Into* [28]. At the heart of this solution is the following very beautiful idea from geometry: if you hold a $2 \times 2 \times 2$ cube so that one of its main diagonals is vertical and then cut the cube exactly in half with a horizontal slice, the cross-section you get will be a hexagon. In fact, the sides of this hexagon are nothing but the diagonals of the six 1×1 squares bisected by the horizontal slice, as shown in Figure 1.17.

Now, here is the nice surprise. This hexagon just happens to form a small knight's mini-tour, which has been labeled $A_1, A_2, \ldots, A_6$ in Figure 1.17. Furthermore, since a cube has four main diagonals, we can in this way construct four distinct hexagons and, hence, four distinct knight's mini-tours. The remaining three mini-tours have been labeled $B_1, B_2, \ldots, B_6$; $C_1, C_2, \ldots, C_6$; and $D_1, D_2, \ldots, D_6$ in Figure 1.17. These mini-tours are, of course, closed 6-cycles which must be joined together in some way. But, note that A_6–B_1, B_6–C_1, C_6–D_1, and D_6–A_1 are all legal knight's moves on the $2 \times 2 \times 2$ cube. Thus, we can form the following knight's tour of the entire $2 \times 2 \times 2$ cube:

$$A_1, A_2, \ldots, A_6, B_1, B_2, \ldots, B_6, C_1, C_2, \ldots, C_6, D_1, D_2, \ldots, D_6, A_1.$$

Solution 1.4 The 8×8 board can easily be covered by either sixteen L tetrominoes or by sixteen T tetrominoes; the simplest ways to do this are shown in Figure 1.18.

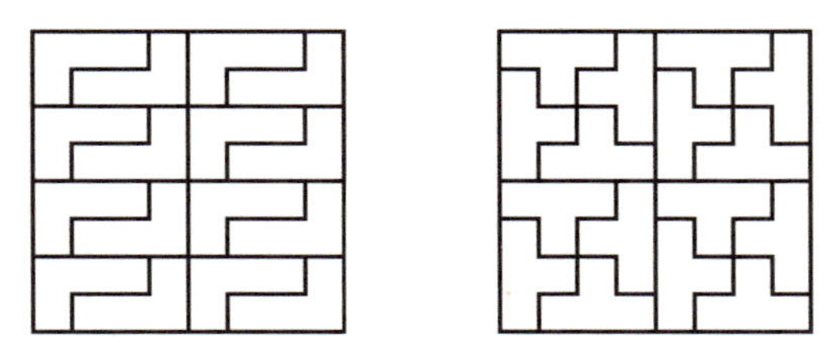

Figure 1.18 Covering the 8×8 board with tetrominoes.

On the other hand, it is impossible to cover the entire 8×8 board using only Z tetrominoes. Suppose you were to start by covering the upper left-hand corner square with a Z tetromino placed horizontally. This would then force you to place another one right next to it, also along the top row, and then a third one is forced as well. But then you are really completely stuck, since there is no way to cover the remaining two squares at the end of the top row. Of course, it makes no difference if you begin in the upper left-hand corner by orienting the first tetromino vertically. The same difficulty arises along the left edge.

It is also impossible to cover either a 4×5 or a 2×10 board with the five distinct polyominoes. A checkerboard coloring of either of these boards has ten white squares and ten black squares. Now, it is easy to see that no matter how you place them on the board, the square tetromino, the straight tetromino, the L tetromino, and the Z tetromino each cover exactly two white squares and two black squares. However, the T tetromino always covers three squares of one color and one square of the other color. So, it is impossible for all five tetrominoes to cover an equal number of white squares and black squares.

Solution 1.5 The solutions for this problem are given in Figure 1.19.

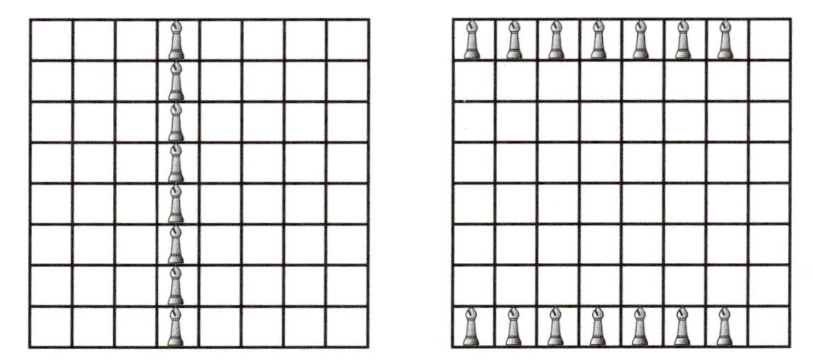

Figure 1.19 Eight dominating bishops and fourteen independent bishops.

Solution 1.6 The solution for this problem is given in Figure 1.20.

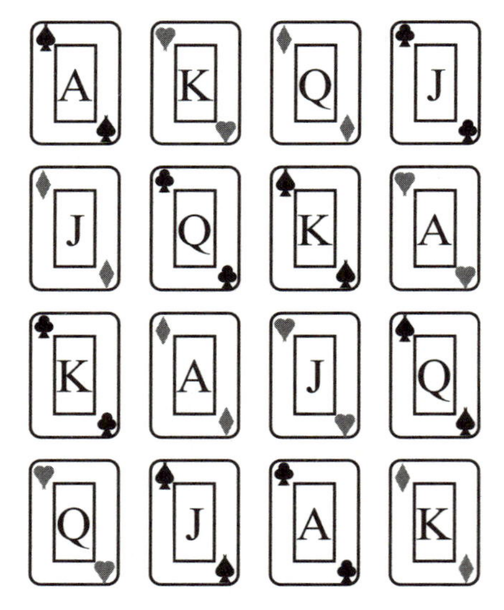

Figure 1.20 A solution to Ozanam's Problem.

CHAPTER TWO

Knight's Tours

A Night at the Theater

There I was, sitting in my seat at the *Second Stage Theatre*, a seat I had paid $80 for mind you, watching renowned sleight-of-hand artist Ricky Jay do a knight's tour on stage in front of a sold-out crowd of several hundred theater goers each of whom had also shelled out roughly $80 for the privilege. I could hardly believe my eyes, it was the summer of 2002, roughly 1000 years after the idea of a knight's tour first emerged somewhere in India, and here I was watching the world's most famous conman—you've probably seen him in such wonderful films as *House of Games* and *Heist*—do a knight's tour on a Broadway stage. Well, OK, it was *off*-Broadway, but it was still very hard to get a ticket for this production of *Ricky Jay On The Stem*, especially since it was being directed by David Mamet. 'The Stem' is an old slang term for Broadway, and this one-man show was itself a delightful tour of the magic, sleight-of-hand, and confidence games that have long been practiced along that famed New York street. I came to the theater that night expecting to be dazzled by many things, but I certainly didn't expect knight's tours.

Of course I knew a bit about this particular con and, in some sense, there really wasn't much to it: Ricky Jay had just memorized a particular knight's tour, that's all. Nonetheless, the excitement in the theater was palpable as Ricky, being blindfolded, called out the next move to his assistant, a young volunteer Ricky had selected from the audience. This volunteer then turned on a very bright light in the center of the square that was to be the next square visited along the knight's path and this was all taking place on a huge vertical chessboard that was easily visible from the very back of the theater. The tension mounted even higher as the tour neared the end: would the blindfolded Ricky be able to find the empty squares that we could see so easily? It did seem almost magical when he called out the last few remaining moves that allowed the knight to complete the tour.

This may sound pretty impressive by itself, and it was, but what you need to visualize is that Ricky Jay was also throwing in several distractions in addition to the aforementioned blindfold that made this already seemingly difficult task of doing a knight's tour all the more impossible. After every few moves of the knight he would abruptly interrupt the tour and do something completely wacky. For example, he had selected another volunteer from the audience and her job was to randomly read to him a six-digit integer and then he would, after a few seconds of intense concentration, produce the cube root of this number! Or, as another distraction, he had previously elicited from the audience, seemingly also at random, the title of one of Shakespeare's plays, this particular evening it was *A Midsummer Night's Dream*, and so, during one of these abrupt pauses in the knight's tour he would suddenly out of nowhere begin reciting a passage from this play, such as: "If we shadows have offended, Think but this, and all is mended, That you have but slumbered here While these visions did appear." The wackiest, though, of these self-imposed interruptions was what seemed to me to be a pretty accurate vocal rendition of a southern mountain holler. But after each of these interruptions, which of course had the audience howling with laughter, he would instantly pick up the tour where he left off. It was not only fun, it was very impressive.

One trick that Ricky used to good effect was that he let the first volunteer, the one who ran the light bulbs, choose the starting square for the knight's tour. This appeared to make his task much harder since he couldn't possibly memorize 64 different tours. In other words, it certainly seems to the audience that it should make a difference where you start the tour. But of course all Ricky has to do is have a single closed tour memorized and he can just enter this one memorized loop at whatever starting point the assistant happens to pick for him. He can memorize one closed tour, and then produce 64 'different' open tours on demand.

I learned something very important that night, something that was well worth my eighty bucks. Knight's tours are inherently dramatic. As the tour nears the end, the options dwindle and it becomes increasingly unlikely that the knight can continue to move. This is something that anyone who tries to do tours on chessboards inevitably feels, and something that the audience also sensed. One result of this dramatic tension is that there is a genuine rush of pleasure when you see before you, almost miraculously, an open path to the end. What is so remarkable to me is that Ricky Jay could take this ancient mathematical puzzle and package it in such a way that he had a New York audience on the edges of their seats, collectively wanting to shout out the next move to a struggling blindfolded man on the stage, and then wildly cheering when Ricky, on his own, somehow managed to call out the correct move to his assistant, waiting patiently all the while to flip the next switch.

Early Work on Knight's Tours

Among the earliest known solutions to the Knight's Tour Problem is an open tour of an 8×8 board by de Moivre which he sent to Brook Taylor early in the eighteenth century. This open tour, shown in Figure 2.1, illustrates a technique that is simple, but surprisingly effective. Start in the outer ring of squares and simply cycle around the outside of the board venturing inside only when forced to do so. In this particular case, the first 24 moves are in the outer two rows. At square 25 you venture inside, but then immediately return to the outer ring at square 26 and continue cycling.

Problem 2.1 Find a knight's tour of a 6×6 chessboard using the de Moivre technique of staying on the outside ring. Take a hint from Ricky Jay: since this is to be a closed tour, it doesn't matter where you start, so you might as well start in a corner.

34	49	22	11	36	39	24	1
21	10	35	50	23	12	37	40
48	33	62	57	38	25	2	13
9	20	51	54	63	60	41	26
32	47	58	61	56	53	14	3
19	8	55	52	59	64	27	42
46	31	6	17	44	29	4	15
7	18	45	30	5	16	43	28

Figure 2.1 An open tour by de Moivre.

Note, however, that in de Moivre's tour in Figure 2.1 a closed tour was made impossible as soon as he landed on square 12, because, as we saw in our discussion of the 4×4 chessboard in Chapter 1, once squares 2 and 12 have been visited there is no longer any way to return to square 1. So, in your application of de Moivre's general strategy you should also pay special attention to corners and their two connecting squares.

At this point I'd like to take the time to look at a single, and rather long, example—one cited by Legendre as being of exceptional difficulty—that was worked out by Euler in 1759. This example was first shown to me by a student, David Carlson, who came across it in W. W. Rouse Ball's *Mathematical Recreations and Essays* [2]. It serves to illustrate the industriousness with which one of the greatest mathematical minds of all time attacked the Knight's Tour Problem. In this example Euler provides us with a powerful and flexible technique that will prove useful to us later over and over again in a variety of situations.

We begin his example somewhat arbitrarily, as one easily might, with a long open tour as shown in Figure 2.2 that misses only four squares: a, b, c, d. Rather than backtracking at this point and trying different paths either at random or one by one, since neither approach seems at all promising, Euler's strategy is to deal with what is missing from our current tour methodically one step at a time. First, he closes the path (ignoring for

55	58	29	40	27	44	19	22
60	39	56	43	30	21	26	45
57	54	59	28	41	18	23	20
38	51	42	31	8	25	46	17
53	32	37	a	47	16	9	24
50	3	52	33	36	7	12	15
1	34	5	48	b	14	c	10
4	49	2	35	6	11	d	13

Figure 2.2 An incomplete open tour.

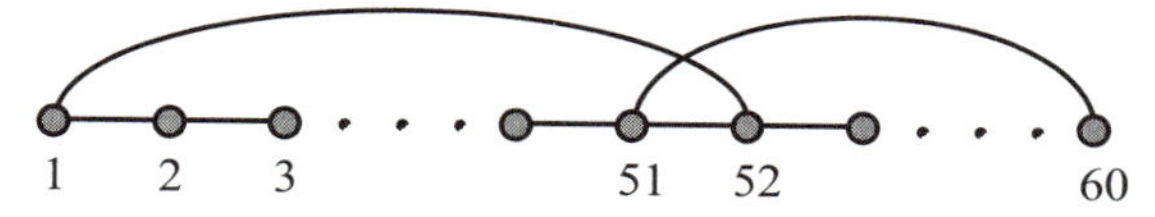

Figure 2.3 Closing the tour.

the time being the four missing squares a, b, c, and d). Since square 1 is adjacent to square 52 while square 60 is adjacent to square 51, it is easy to see from Figure 2.3 that we can form a closed tour of these 60 squares by merely tracing the original tour from 1 to 51, then jumping to 60 and working backwards to 52 and finally jumping back to 1. This closed tour, appropriately renumbered, is recorded in Figure 2.4.

Next, Euler attacks the problem of bringing the four missing squares into the tour. Since three squares—a, b, and d—happen to form a path, we can attach this path to our already existing 60-cycle (and we just continue to ignore c for the time being). I should mention that there are several other attractive alternatives here. For example, it would also be easy to insert a and b between 2 and 3 as in 2-b-a-3 or to insert b and d between 15 and 16 as in 15-d-b-16. But, staying with Euler, we simply attach a (and hence b and d) to one of the squares to which it is adjacent, say, 51 (see Figure 2.5). By then breaking the edge between 51 and 52, we create an open tour beginning at

57	54	29	40	27	44	19	22
52	39	56	43	30	21	26	45
55	58	53	28	41	18	23	20
38	51	42	31	8	25	46	17
59	32	37	a	47	16	9	24
50	3	60	33	36	7	12	15
1	34	5	48	b	14	c	10
4	49	2	35	6	11	d	13

Figure 2.4 A closed tour.

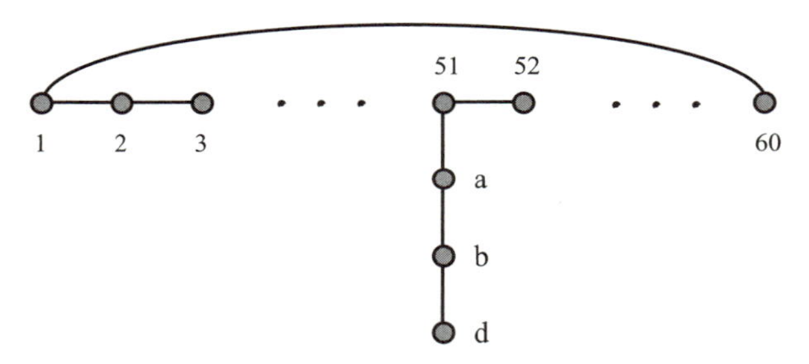

Figure 2.5 Extending the tour.

52, which we now renumber 1, and ending at d, which is now renumbered 63. This open tour is shown in Figure 2.6.

We still need to include square c and, of course, re-close the tour. But, since c is adjacent to 25 while 63 is adjacent to 24, we can visit all 64 squares by beginning at 1, continuing on to 24, jumping to 63, going backwards to 25, and then ending at c, as shown in Figure 2.7. This tour, again appropriately renumbered, is recorded in Figure 2.8.

Finally, we must close the tour. The main difficulty is that 1 and 64 are rather far apart. To handle this difficulty, Euler chooses a square such as 28, which is adjacent to 1 and which has a neighbor 27 that is close to 64 (see Figure 2.9). We can thus begin a tour at 27 and work backwards to 1, return to 28 and continue on to 64. This renumbered tour is shown in Figure 2.10.

6	3	38	49	36	53	28	31
1	48	5	52	39	30	35	54
4	7	2	37	50	27	32	29
47	60	51	40	17	34	55	26
8	41	46	61	56	25	18	33
59	12	9	42	45	16	21	24
10	43	14	57	62	23	c	19
13	58	11	44	15	20	63	22

Figure 2.6 An extended tour.

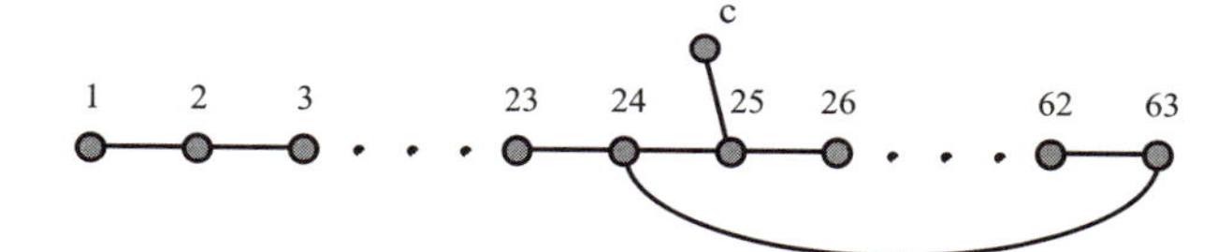

Figure 2.7 Including c in the tour.

6	3	50	39	52	35	60	57
1	40	5	36	49	58	53	34
4	7	2	51	38	61	56	59
41	28	37	48	17	54	33	62
8	47	42	27	32	63	18	55
29	12	9	46	43	16	21	24
10	45	14	31	26	23	64	19
13	30	11	44	15	20	25	22

Figure 2.8 A complete open tour.

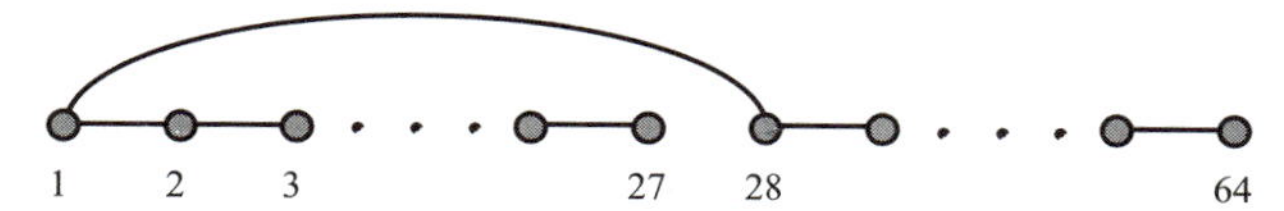

Figure 2.9 Getting the ends closer together.

22	25	50	39	52	35	60	57
27	40	23	36	49	58	53	34
24	21	26	51	38	61	56	59
41	28	37	48	11	54	33	62
20	47	42	1	32	63	10	55
29	16	19	46	43	12	7	4
18	45	14	31	2	5	64	9
15	30	17	44	13	8	3	6

Figure 2.10 Squares 1 and 64 are now close enough.

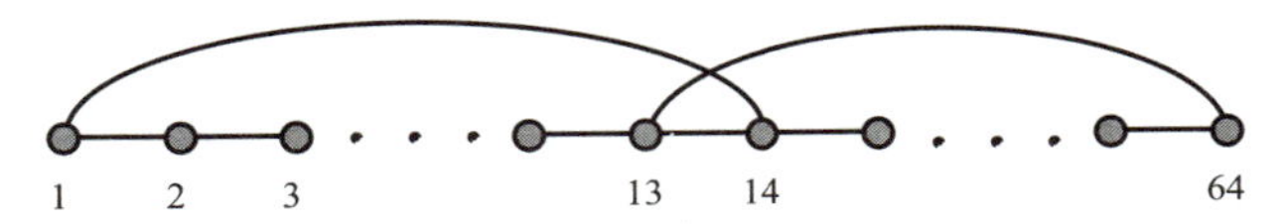

Figure 2.11 Closing the tour.

22	25	50	39	52	35	60	57
27	40	23	36	49	58	53	34
24	21	26	51	38	61	56	59
41	28	37	48	3	54	33	62
20	47	42	13	32	63	4	55
29	16	19	46	43	2	7	10
18	45	14	31	12	9	64	5
15	30	17	44	1	6	11	8

Figure 2.12 The final tour.

Now, since 1 is adjacent to 14 while 64 is adjacent to 13 (see Figure 2.10), we see in Figure 2.11 how we can now close the tour by beginning at 13, going backwards to 1, jumping to 14 and continuing on to 64 and then directly back to 13. This final tour is given in Figure 2.12.

You might perhaps come away from this example with the feeling that Euler was just lucky, that for instance it was a bit too convenient at the end that 1 and 14 were adjacent and so were 64 and 13. But, trust me, in practice, this sort of thing either works out or can be made to work out. The point is that there were lots of possible combinations that were favorable: 1 and 3 along with 64 and 2, 1 and 4 with 64 and 3, and so on. So the odds were pretty good—in fact, about 1 in 3—that two of these would both turn out to be adjacent knight moves. So if that sort of thing does work out, great; if not, you just try something else until you do get lucky. In other words, Euler wasn't lucky, he was just persistent.

Problem 2.2 If you look carefully at Figure 1.6 in Chapter 1, you will see that the graph for the 4×4 chessboard actually contains within it four disjoint 4-cycles: the bold diamond-shaped 4-cycle, another diamond-shaped 4-cycle slanted in the other direction, as well as two square 4-cycles that are slightly tipped. Now, think of this 4×4 graph—and, more importantly, these four 4-cycles—as being replicated in each of the four quadrants of the 8×8 chessboard. Then, imitating Euler, create a knight's tour of the entire 8×8 board by linking these 4-cycles together into one big tour. You might want to start by linking the four 4-cycles of a given type together first, such as the four that correspond to the bold diamond-shaped 4-cycle shown in Figure 1.6.

Exactly 100 years after Euler published his own contribution to the Knight's Tour Problem, the great Irish mathematician, Sir William Rowan Hamilton, marketed (in 1859) a board game called the *Icosian Game* in which the object was to complete a tour of the vertices of the graph of the dodecahedron having already been given the first five vertices to be visited. There was even a deluxe edition of the game, *A Voyage Round the World*, in which the 'board' was a solid dodecahedron and the vertices of the dodecahedron represented exotic locations such as Delhi and Zanzibar. Because of this game, any closed tour in a graph that visits each vertex of the graph exactly once is now called a *Hamiltonian cycle*. In other words, when we are looking for

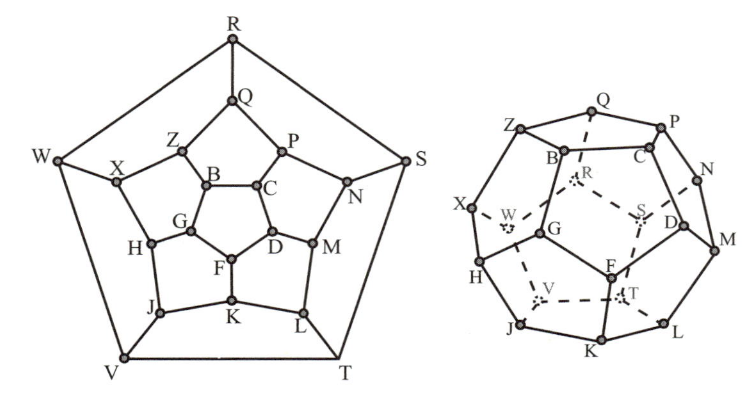

Figure 2.13 Hamilton's *Icosian Game.*

a knight's tour on a chessboard we are in actuality trying to find a Hamiltonian cycle in the associated graph. Similarly, the rook's tour we used to prove Gomory's Theorem in Chapter 1 was nothing more nor less than a Hamiltonian cycle.

Problem 2.3 Figure 2.13 shows the 'board' for the *Icosian Game*, that is, a graph of the dodecahedron. Complete a Hamiltonian cycle that begins with the five vertices B, C, P, N, and M.

Chessboards can certainly be toured by chess pieces other than the knight. For example, in Chapter 1 we made use of a rook's tour to prove Gomory's Theorem. Still, for obvious reasons, the knight is the only chess piece for which mathematical questions about touring have generated much general interest. Not surprisingly, however, there are numerous recreational puzzles involving tours by other pieces. Here are three samples involving the bishop, the queen, and the king. It will frequently be convenient for us to use an ordinary coordinate system for naming the squares on a chessboard. So, our standard convention will be that $(1, 1)$ will represent the lower left-hand corner square and $(8, 8)$ will be the upper right-hand corner.

Problem 2.4 Can a black bishop start at the black square $(1, 1)$, visit each black square of the chessboard along each diagonal exactly once, and end at the black square $(8, 8)$? This question probably needs some clarification. In other words, if a bishop passes through a square along one diagonal, it can pass through

Figure 2.14 A king's tour.

the same square again later along the other diagonal, whereas the other option is for a bishop to enter a square along one diagonal, turn, and then leave along the other diagonal, in which case, the bishop may not visit this square again. This actually makes sense as a tour because the bishop is using each diagonal of each square exactly once. (So, perhaps there is a Bishop's Tour Problem after all!)

Problem 2.5 The famous British puzzle master H. E. Dudeney posed this lovely problem in 1906: move a queen from square $(3, 3)$ to square $(6, 6)$ in 15 moves, visiting every square exactly once and without letting the queen cross her own path.

Problem 2.6 This unusual problem was communicated to me by a student, Kagen Schaefer, who is now a maker of astonishingly beautiful and mathematically intricate puzzle boxes. A king's tour of a chessboard inevitably involves horizontal and vertical moves. For example, in the tour shown in Figure 2.14 the king makes 14 horizontal moves and 16 vertical moves. Show that in any king's tour of an 8×8 chessboard in which the king does not cross his own path the sum of the horizontal and vertical moves must be at least 28.

Solutions to Problems

Solution 2.1 A solution is shown in Figure 2.15. We begin in the corner at 1, but then we must be careful to protect access back to the corner at square 1, so as we cycle around the outside this forces us to venture inside to square 8. We can return to the outer ring at 9 and everything is fine, dipping inside again at 17, and then once more at square 26. At this stage it is best to take stock of the situation. Rather astonishingly, however, of the four available moves from square 26, three of these choices lead easily to a complete tour.

34	7	24	15	32	1
23	14	33	36	25	16
6	35	8	17	2	31
13	22	29	26	9	18
28	5	20	11	30	3
21	12	27	4	19	10

Figure 2.15 A 6×6 tour.

Solution 2.2 A solution is shown in Figure 2.16.

11	62	27	48	13	50	31	34
26	47	12	63	30	33	14	51
61	10	45	28	49	16	35	32
46	25	64	9	36	29	52	15
7	60	21	44	1	56	17	38
24	43	8	57	20	37	2	53
59	6	41	22	55	4	39	18
42	23	58	5	40	19	54	3

Figure 2.16 An 8×8 tour.

Solution 2.3 Note that you must follow M with D, otherwise if you follow instead with L you will never be able to return to D. By being careful to never inadvertently isolate a vertex in this way, it is relatively easy to conclude that there are only two ways to complete this cycle, namely,

BCPNMDFKLTSRQZXWVJHGB

and

BCPNMDFGHXWVJKLTSRQZB.

Solution 2.4 A solution is shown in Figure 2.17.

Figure 2.17 A bishop's 'tour'.

Solution 2.5 A solution to Dudeney's problem is shown in Figure 2.18.

Solution 2.6 There are two key observations to be made: the first is that if we imagine the king tracing a curve as he moves along his path, then this simple closed curve will separate the plane into two parts, one part inside the curve and one part outside the curve; thus, the king must visit the perimeter squares of the chessboard *in order*; otherwise, intermediate perimeter squares would become isolated. The second observation to be made is that when a king makes a diagonal move he stays on

Figure 2.18 A queen's tour.

Figure 2.19 A king's tour with fourteen horizontal moves and fourteen vertical moves.

squares of the same color, like a bishop, and only by making a vertical or horizontal move does he change the color of the square he is on. Now, since there are 28 perimeter squares and they have to be visited in order and they alternate between black and white, it follows that there must be at least one horizontal or vertical move used by the king in getting from one perimeter square to the next, and so at least 28 such moves are required. Figure 2.19 exhibits a king's tour in which the number 28 is actually achieved.

CHAPTER THREE

The Knight's Tour Problem

Impossible Tours

In Chapter 1 we saw that a knight's tour is impossible on the 4×4 chessboard and also on any board both of whose dimensions are odd. Knight's tours are also impossible for a number of other chessboards. For example, if a board has only one or two rows, there is not enough room for a tour; with one row, a knight can't even move, and with two rows, a knight can only move in one direction but is then stuck at that far end of the board. By the way, I should mention that, in general, we will usually orient our chessboards horizontally, so that for an $m \times n$ chessboard with m rows and n columns, we have $m \leqslant n$.

What about boards with three rows? In Problem 1.2 in Chapter 1 I mentioned that a 3×10 board was the smallest such board for which a knight's tour is possible. So let's begin this section by seeing why there are no tours for the 3×6 or the 3×8 boards. The 3×4 board, which also has no tour, we will leave for later as a special case of another situation, and so we will ignore it for now.

The 3×6 *Chessboard*

For the 3×6 board we will use an easy idea from graph theory. I'll describe the idea first in terms of necklaces. Suppose you have in your hands a simple single-strand gold chain necklace: if you remove one link from the necklace, then the necklace falls open but remains in one piece; if you remove two links, however, the necklace will fall into two pieces (or perhaps one piece

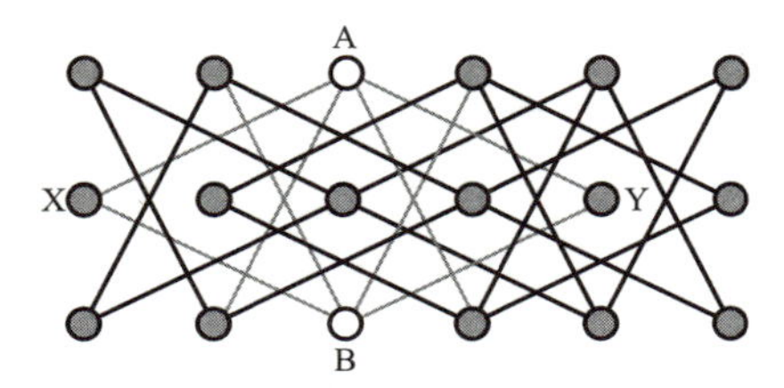

Figure 3.1 Removing two vertices from the knights graph of the 3×6 board.

if the two links happened to be right next to one another); three links, three pieces; and so on. In other words, if you remove k links, the necklace will be in k, or possibly fewer, pieces. But the main point is that the necklace will *never* end up in more than k pieces if you remove k links.

We can use this idea to show that the knights graph for the 3×6 chessboard does not have a Hamiltonian cycle, that is, there is no knight's tour. In Figure 3.1 we see what happens when we remove, or delete, the two vertices A and B from the knights graph for the 3×6 board. Note that, in a graph, when you delete a vertex you must obviously also remove any edges attached to that vertex. Here, the graph has fallen into three pieces: vertex X is isolated as one piece, vertex Y is isolated as the second piece, and the rest of the graph is still connected together into a third piece. So, we removed two vertices, but we ended up with three pieces, which means that this graph could not possibly have had a Hamiltonian cycle. Thus, the 3×6 board does not have a knight's tour.

Problem 3.1 In Figure 3.2 the squares of the 4×4 chessboard have been labeled in a way that brings out the underlying structure of the unfolded knights graph. Use this graph to show that the 4×4 chessboard does not have a knight's tour by finding k vertices in the graph whose removal causes the graph to fall into more than k pieces, making a Hamiltonian cycle impossible in this graph. Don't forget that when you delete a vertex, you must also delete the edges attached to it.

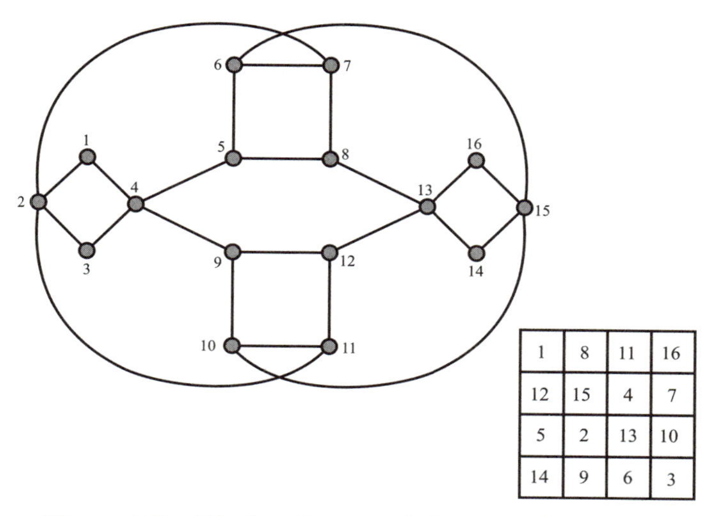

1	8	11	16
12	15	4	7
5	2	13	10
14	9	6	3

Figure 3.2 The knights graph for the 4×4 board.

The 3×8 *Chessboard*

In graph theory, we refer to the number of edges attached to a vertex as the *degree* of the vertex. It is clear—and also an idea we have used before—that if a graph has a Hamiltonian cycle, then, for any vertex of degree 2 in the graph, both of the edges attached to that vertex must be a part of the Hamiltonian cycle. After all, that's the only way the Hamiltonian cycle can pass through a vertex of degree 2.

Now, the knights graph for the 3×8 board has precisely eight vertices of degree 2. So let's assume for the moment that this graph does have a Hamiltonian cycle; then, in Figure 3.3 on the left, we show all of the edges that must be a part of that Hamiltonian cycle simply because they are attached to one of the eight vertices of degree 2. These eight vertices have been shaded in the figure to make them easy for you to spot.

Note that there are two isolated vertices in the middle, X and Y, which must also be included in the Hamiltonian cycle. How are these two vertices to be connected into the rest of this cycle? Clearly, one edge leading to X must come from the left and one edge must come from the right; otherwise, with two edges connecting X to the same side of the graph you automatically create

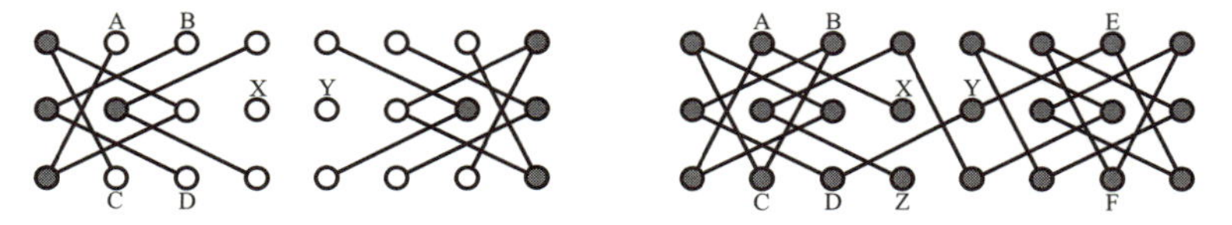

Figure 3.3 The knights graph for the 3×8 board.

a small cycle on that side, thereby making a larger Hamiltonian cycle impossible. The same is true for Y, one edge must come from the left and one edge from the right. Therefore, by symmetry, we may as well assume that the edge AX is part of the Hamiltonian cycle. Now, let's follow the cycle from X to A and see where it goes. Continue past A from edge to edge until you get to C. From C, the Hamiltonian cycle can't go back to X yet, so it must go to B next, from where we follow it on to D. At D there again seems to be a choice of which way to go, but remember that there must be an edge attached to vertex Y coming from the left, so our Hamiltonian cycle must now go from D to Y. Since the remaining situation is now once again symmetric, we may as well assume that the Hamiltonian cycle continues next from Y on to E, rather than F, and from there we follow the cycle along until we reach vertex Z and find ourselves unable to continue. This dead end means that there never was any Hamiltonian cycle after all. So, alas, the 3×8 chessboard has no knight's tour.

The $4 \times n$ *Chessboards*

It is rather amazing, but we have already seen almost all of the chessboards that have no knight's tours. There remains only one additional category of impossible boards for us to consider. This is the category of boards that have four rows. This may seem strange and, in fact, it *is* strange, because there is plenty of room for knights to maneuver on these boards, and yet these boards have no tours. The only hint we have had so far about this strange state of affairs is that of course we know that the 4×4 board doesn't have a knight's tour; and it has also been hinted that neither does the 4×3 board. So, as it turns out, it isn't the smallness of these boards that is the problem, it is the fact that they have four rows. Any board with four rows can't have a knight's tour. Period.

I am going to present two proofs of the impossibility of knight's tours for $4 \times n$ boards. Both proofs use coloring and both are very clever. The first proof is by Louis Pósa, a child prodigy from Hungary, who made a number of significant mathematical discoveries while still a very young teenager. The second proof is by Solomon W. Golomb, whom we will meet again later in the book as the inventor of polyominoes.

Pósa's Proof

Suppose that, for a given value of n, the $4 \times n$ chessboard does in fact have a knight's tour. With the standard black and white checkerboard coloring of this $4 \times n$ board, we know that this knight's tour must alternate between the white squares and the black squares. Now, imagine that we have re-colored this chessboard so that the entire top row and the entire bottom row of squares are colored red and that the two middle rows are entirely blue. Let's look at the knight's tour again. Since a knight on a red square can only move to a blue square, every time a knight is at a red square during the knight's tour it must visit a blue square next on the tour. Moreover, since there are the same number of red squares and blue squares to be visited, a knight can't dare visit two blue squares in a row anywhere along the tour since it wouldn't be able to make up for this later by visiting two red squares in a row. Therefore, we conclude that the knight's tour must strictly alternate between the red and the blue squares. But this is impossible, the same knight's tour alternated between white and black squares in the one coloring, and between red and blue squares in the other coloring, which would mean the two color patterns must be the same, which of course they aren't. Isn't that a clever proof, especially for a teenager to discover?

Golomb's Proof

Again, suppose that, for a given value of n, the $4 \times n$ chessboard does in fact have a knight's tour. Now imagine that we

A	B	A	B	A	B	A
C	D	C	D	C	D	C
D	C	D	C	D	C	D
B	A	B	A	B	A	B

Figure 3.4 Golomb's coloring for the $4 \times n$ board.

have colored this chessboard using four colors, A, B, C, and D, in the checkerboard pattern shown in Figure 3.4. Don't worry about how the pattern ends on the right; the exact number of columns in the figure turns out to be irrelevant to the discussion. Note that a knight at a square colored A can only move to a square colored C, so when a knight visits a square colored A while traveling along the knight's tour, the knight must have just come from a square colored C and be just about to go to still another square colored C next on the tour. Moreover, since there are an equal number of squares colored A and squares colored C, when a knight moves from a square colored A to a square colored C, it can't then go on from the square colored C to a square colored D, for the knight would still have to eventually return to another square colored C before being able to get back to a square colored A. In other words, the knight has to avoid squares colored B and D entirely in order to visit all of the squares colored A. Therefore, no knight's tour is possible. Incidentally, this color pattern was also used somewhat earlier by Maurice Kraitchik to show that a knight's tour is impossible for the $4 \times n$ board, but his argument was considerably more involved [20].

Schwenk's Theorem

At this point we really have seen all of the chessboards that don't have a knight's tour. All other rectangular boards have tours! Given the extraordinary attention that the Knight's Tour Problem has received over the past several centuries it is quite surprising that a definitive solution to this problem didn't appear in print until a beautiful paper by Allen Schwenk was published in *Mathematics Magazine* in 1991 [27]. This paper pulled together all of the pieces: a full discussion of the well-understood impossible boards, as well as the much more challenging, and usually ignored, task of proving that all other boards have knight's tours. The final story for the Knight's Tour Problem can now be summarized in the following remarkable theorem.

Theorem 3.1 (Schwenk's Theorem) *An $m \times n$ chessboard with $m \leqslant n$ has a knight's tour unless one or more of the following three conditions hold:*

(a) *m and n are both odd;*

(b) *$m = 1, 2,$ or 4; or*

(c) *$m = 3$ and $n = 4, 6,$ or 8.*

So, then, exactly how does one go about proving that all boards, except those explicitly excluded by these three conditions, have knight's tours? We saw one strategy in Chapter 2 that had been used by de Moivre: keep to the outside ring of squares. This worked fairly well on modestly sized chessboards, but it's hard to believe it would keep working on boards when the size of the board gets really big. A far more rigorously algorithmic heuristic from 1823 was used by Warnsdorff in which the knight always moves to a square from which it would control the fewest squares not already visited. You might want to try this 'algorithm' out on, say, the 3×10 chessboard or the 6×6 chessboard. Nevertheless, I think it is fairly clear that we are not likely to be able to find an algorithm like this that would work for all boards; in part, this is because any such algorithm would have to have built into it an ability to reject the excluded boards. So, a somewhat different strategy is needed.

$3 \times n$ *Chessboards*

The approach that works is an inductive one: start with a tour of a small board and slowly but steadily build larger and larger tours from smaller tours. This is a very pretty idea. Let's see how the idea works in a specific case by showing that any $3 \times n$ board has a knight's tour for $n \geqslant 10$ and n even, that is, all of the boards, 3×10, 3×12, 3×14, 3×16, 3×18, ... —this sequence goes on forever—have tours. We start small in Figure 3.5 with tours for the 3×10 and the 3×12 boards. Normally, in an inductive argument like this, we would just start with the smallest board, namely, the 3×10 board, but we actually need to begin with tours of both of these boards in this case, as you will see.

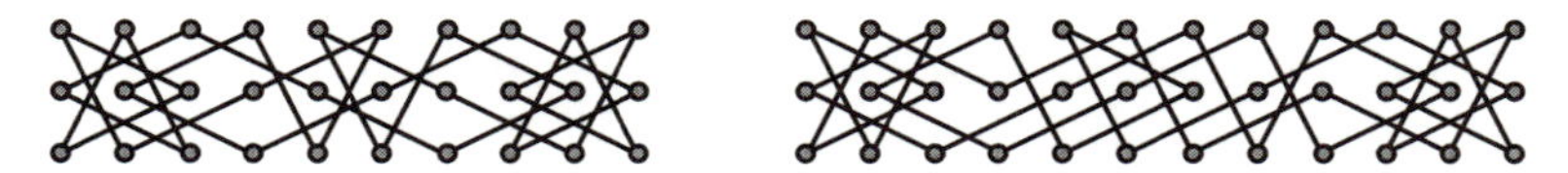

Figure 3.5 Knight's tours for the 3×10 and the 3×12 boards.

This lays the foundation. Next, we have to see how you can add on additional columns and get tours for slightly larger boards. It turns out that we can do this easily by adding four columns at a time. This process is illustrated in Figure 3.6, in which the left part of the diagram is the existing tour, in this case either the 3×10 tour or the 3×12 tour, and the right part of the diagram is an *open* tour of a 3×4 board. Now, we remove a single edge on the left from the existing tour—the dotted edge—and then join the two tours together with two bold edges to get a single closed tour of the larger board; in this case, that is, we now have a tour of either a 3×14 board or a 3×16 board. We can now clearly repeat this procedure exactly to get a tour for a 3×18 and a 3×20 board. And so on, forever. Note that because we add four columns each time it was sufficient for us to start with just the two boards having 10 and 12 columns, respectively, in order eventually to achieve tours for all larger boards with an even number of columns.

So that's the main idea behind Schwenk's proof. What he is able to show—and it is not as hard as it is going to sound—is that you can start with tours for just nine small boards: the 3×10 and 3×12 we have just discussed, as well as tours for the 5×5, 6×6, 5×8, 6×7, 6×8, 7×8, and 8×8 boards, and then from these small beginnings you can slowly build tours for any boards not explicitly excluded by the three conditions of his theorem.

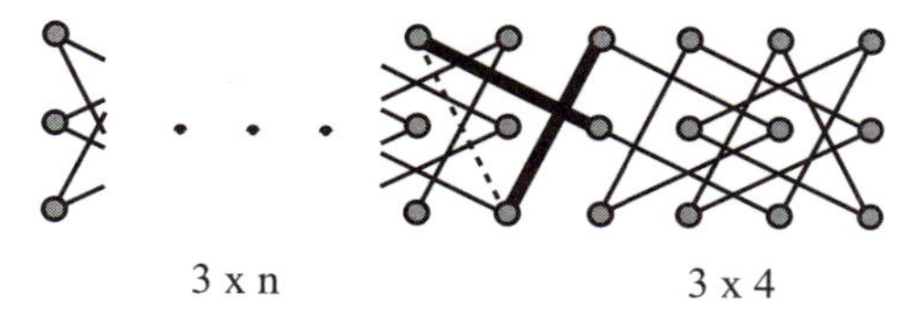

Figure 3.6 Adding four new columns to a $3 \times n$ tour.

Problem 3.2 The logo that See's Candies uses for all of its products has a checkerboard border consisting of just the outer three-row ring from a rectangular chessboard. Figure 3.7 shows the design on one of their shopping bags. Does this 37×57 annular 'chessboard' have a knight's tour? What about a similar 23×44 annular 'chessboard' that appears as a border on one of their boxes of candies? Does it have a tour?

Figure 3.7 Does this 'chessboard' have a knight's tour?

Problem 3.3 Find an open knight's tour for the 5×5 chessboard that begins in the center square of the board and ends in a corner square. Next, show how to extend this open tour to an open knight's tour for a 9×9 chessboard by adding an outer two-row ring to a 5×5 board; then repeat this process to get an open tour for a 13×13 board, and so on. In this way, therefore, show that any $(4n+1) \times (4n+1)$ chessboard has an open knight's tour.

Problem 3.4 Does the 4×4 chessboard have an open knight's tour?

Problem 3.5 Here is a slightly unusual problem from Martin Gardner that at first may not seem to have much to do with knight's tours [14]. Suppose each square of a 5×5 chessboard is occupied by a knight. Is it possible for all 25 knights to move simultaneously? The fact that a 5×5 chessboard doesn't have a knight's tour is what makes this problem interesting, because if you ask the same question about a board that does have a knight's tour then the answer is far too easy because all of the knight's can simply take one step forward along the knight's tour all at the same time.

Solutions to Problems

Solution 3.1 If you remove the four vertices along the middle row, that is, vertices 2, 4, 13, and 15, then you will be left with six components: four isolated vertices and two squares. Therefore, the graph has no Hamiltonian cycle, and no knight's tour is possible for the 4×4 board.

Solution 3.2 Yes, both of these 'chessboards' have tours. We can decompose the 37×57 'chessboard' into two 3×54 chessboards and two 3×34 chessboards as shown in Figure 3.8. By Schwenk's Theorem, we know that all four of these smaller chessboards have knight's tours. These four smaller tours can then be joined into a single tour of the entire board, as indicated in Figure 3.8, by removing the six dotted edges—these edges are guaranteed to be part of the four original tours simply because they are attached to vertices of degree 2—and replacing them as indicated with the six solid edges.

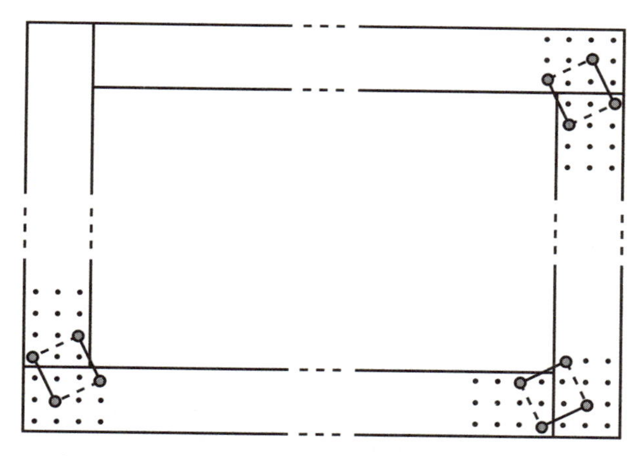

Figure 3.8 A knight's tour for the 37×57 'chessboard'.

The 23×44 'chessboard' can be done in exactly the same way, but Figure 3.9 shows a much more satisfying solution found by a student, Yulan Qing.

Figure 3.9 A knight's tour for the 23×44 'chessboard'.

Solution 3.3 The required open knight's tour for the 5×5 board is shown in the center of Figure 3.10. The surrounding two-row ring also contains an open tour found by starting one square above the lower left-hand corner and simply cycling clockwise around the ring until you come to an end in the corner, quite amazed at how well it all worked out. This can be repeated for any number of two-row rings. The bold edges then connect these individual open tours into a tour of the entire board. In fact, you can even think of this process as having started in the center with a knight's tour of the 1×1 chessboard!

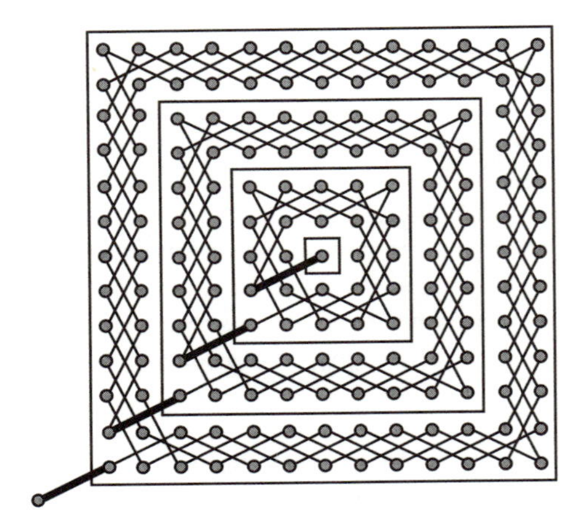

Figure 3.10 Open knight's tours for the $(4n + 1) \times (4n + 1)$ chessboards.

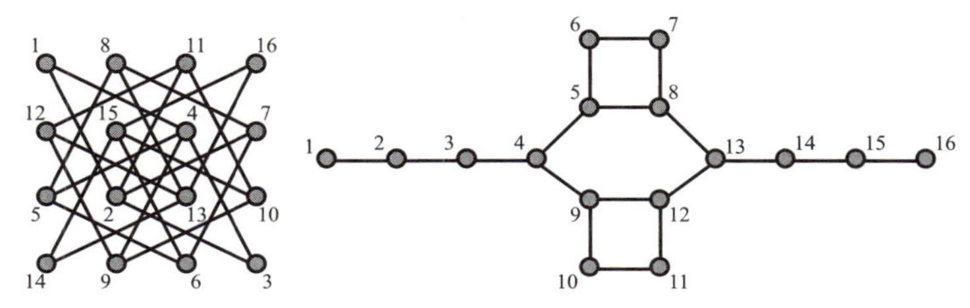

Figure 3.11 An attempt at an open tour of the 4×4 chessboard.

Solution 3.4 No, not even an open tour is possible. Label the knights graph as in Figure 3.11. Earlier we argued that the 4×4 board doesn't have a closed knight's tour because vertices 1 and 3 each have degree 2, which would force the four edges of a 4-cycle, namely, edges 1–2, 2–3, 3–4, and 4–1, to be in any Hamiltonian cycle of the entire graph, which is clearly impossible. The same argument, of course, could have been applied to vertices 14 and 16, since they also each have degree 2. But, now, if we consider the possibility of an open Hamiltonian path, that is, a path passing through every vertex exactly once but not returning to the starting vertex, then, by the same reasoning, such a Hamiltonian path must start at either vertex 1 or vertex 3, and must end at either vertex 14 or vertex 16.

By symmetry, we will assume that this Hamiltonian path begins at vertex 1 and we may as well also assume that the path then goes to vertex 2. It must then go to vertex 3, followed next by 4; otherwise vertex 3 will become isolated. The situation concerning where the Hamiltonian path ends is also symmetric and, for the same reason, we may as well assume that it ends at vertex 16, which means that the end of the Hamiltonian path can be assumed to consist of 13–14–15–16. This is all depicted in Figure 3.11, where it is now clear that whichever way the knight decides to go from vertex 4, either left or right, several vertices will inevitably be missed on the way to vertex 16. Thus, even an open knight's tour is impossible for the 4×4 chessboard.

By the way, you can give an alternative, and far easier, argument based on the idea appearing in Problem 3.1 and in Figure 3.1. Since you can remove four vertices from the knights graph of the 4×4 chessboard and have the graph fall into six pieces, there could not possibly have been a Hamiltonian path (you can't make four cuts in a single length of string and end up with six pieces of string—well, most people can't, maybe Ricky Jay could do it).

Solution 3.5 It is not possible. With the usual checkerboard coloring, there are 13 squares of one color and 12 squares of the other color. Suppose the board has 13 white squares. When the 13 knights on the white squares move they need to all land simultaneously on only 12 black squares, which is impossible.

Magic Squares

Muhammad ibn Muhammad

At least a generation before Euler was doing his own work on knight's tours in Switzerland, there was a Fulani mathematician and astronomer living in the city of Katsina in a region of West Africa that is now the northernmost part of Nigeria, but which at that time was at the southern end of major trade routes that crossed the Sahara to northern Africa. In 1732, this African mathematician, Muhammad ibn Muhammad, wrote a manuscript, in Arabic, concerning the construction of magic squares. Amazingly, this manuscript has survived.

Claudia Zaslavsky described Muhammad ibn Muhammad's work on magic squares in her delightful book on African mathematics, *Africa Counts* [41]. Before I get to the way he made use of the knight's move in order to construct magic squares, let me begin with a simpler construction of his for magic squares of odd order, that is, for squares of size 3×3, 5×5, 7×7, and so on.

Magic Squares of Odd Order

Muhammad ibn Muhammad's construction for magic squares of odd order involved an unusual convention. He thought of the columns of the squares as wrapping from the bottom of each column back to the top, and similarly he also thought of each row as wrapping from its right side back around to its left side. This should sound familiar, because it is exactly what we were doing in Chapter 1 in Figure 1.7 when the knight closed

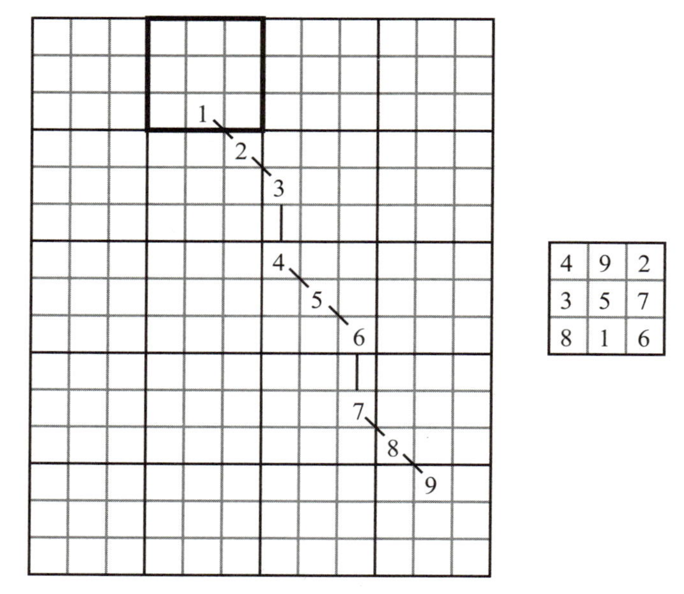

Figure 4.1 Muhammad ibn Muhammad's construction of the Lo-shu magic square.

its open tour of the 5 × 5 chessboard by going off the bottom edge of the board and reappearing at the top.

So, with this convention, here is Muhammad ibn Muhammad's simple construction, stated here for a 3 × 3 square: start with 1 in the square immediately below the center square of the 3 × 3 chessboard, then go diagonally right and down one square for 2, and so on; each time you are blocked—that is, every third move—go straight down two squares instead. This is illustrated in Figure 4.1 showing first just the bare construction without the wrapping, and then the effect of the wrapping to produce the 3 × 3 magic square on the right.

This 3 × 3 array of the integers from 1 to 9 is called a *magic square* because the sum of the numbers in each row, each column, and each of the two main diagonals is exactly the same, namely, 15. That's eight different sums and they are all the same! This particular magic square is a very famous one and is also very old. There is even a legend that over 4000 years ago a turtle emerged from the Yellow (Lo) River in China with this magic square on its shell, and so this pattern is now called the *Lo-shu* magic square.

Problem 4.1 Use Muhammad ibn Muhammad's construction to produce magic squares of order 5×5 and 7×7. For the 5×5 magic square the sum of the numbers in each row, each column, and each of the two main diagonals will be 65; and for the 7×7 magic square the sum will be 175.

It is not at all hard to see why this construction works. Since every time you are blocked you begin the next sequence of numbers in a different starting column, you end up evenly distributing the numbers among the columns. The same is true for the rows, although this is somewhat less obvious because you move down two squares each time. The diagonals are taken care of because Muhammad ibn Muhammad carefully started by placing the 1 so that the average number—that is, 5 for the 3×3 case, 13 for the 5×5 case, and 25 for the 7×7 case—would be placed in the center square of the board and observing that along each main diagonal the numbers get bigger in one direction from the center at the same rate they get smaller in the other direction from the center. This very same construction was discovered by Bachet in the early 1600s and a nearly identical construction was brought back to France from Siam, now Thailand, by de la Loubère.

Using Knights to Construct Magic Squares

Muhammad ibn Muhammad also had a construction for magic squares that used the knight's move. The idea of this construction is really the same as before, but instead of the basic move being a single diagonal step as it was earlier, the basic move is now a knight's move. So, as shown in Figure 4.2, the knight begins in a corner at 1 and makes successive knight's moves—all in the same direction, wrapping as necessary—to 2, 3, 4, and 5. Here, the knight is blocked, since continuing in this same direction would put the knight back at square 1, so at this stage Muhammad ibn Muhammad moves the knight two squares to the left, from 5 to 6, and then resumes the knight's moves in

13	25	7	19	1
17	4	11	23	10
21	8	20	2	14
5	12	24	6	18
9	16	3	15	22

Figure 4.2 Muhammad ibn Muhammad's 5×5 magic square using the knight's move.

the previous direction. Similarly, 11 is placed two squares to the left of 10, and so on. An added bonus with this construction is that all of the diagonals, not just the two main ones, have the same sum.

Problem 4.2 Muhammad ibn Muhammad also gave a variation on the construction using the knight's move where, instead of moving two squares to the left when the knight was blocked, he moved three squares straight up. Use this method to construct a 7×7 magic square. Note that all fourteen diagonals have the same sum.

In 1996—and in complete and blissful ignorance of Muhammad ibn Muhammad's work—a student, Barry Balof, and I discovered an almost identical method for constructing magic squares, a method, however, that not only used the knight's move but used a knight's tour! The only difference in our method is that, instead of moving left or right or up or down when the knight is blocked, we use a knight's move to get unblocked. An example is given in Figure 4.3 for a 7×7 magic square. It was of course quite natural for a writer of Arabic to start in the upper right-hand corner; similarly, it was natural for us to start in the upper left-hand corner. Note that at square 7, where we are blocked, the knight makes a *knight's move* up and to the right, and again does the same thing at 14, and so on.

Having stumbled upon this nice idea somewhat by accident, we were able to prove that this method does work in general and produces an $n \times n$ magic square using a knight's tour as long as n is not divisible by 2, 3, or 5 [3]. Or, if n is not divisible by 2 or 3, but is divisible by 5, then everything works out except that the sums for the two main diagonals are wrong; so, in this

1	24	47	21	37	11	34
12	35	2	25	48	15	38
16	39	13	29	3	26	49
27	43	17	40	14	30	4
31	5	28	44	18	41	8
42	9	32	6	22	45	19
46	20	36	10	33	7	23

Figure 4.3 A knight's tour magic square by Balof and Watkins.

case, you almost get a magic square, an idea we will talk about in the next section.

Problem 4.3 Construct a 13×13 magic square using a knight's tour.

Semi-Magic Squares

The *Holy Grail* in the quest for magic squares is to find an 8×8 magic square using a knight's tour. Several people have come very, very close. In Figure 4.4 we show an 8×8 square found by Euler which is very close to being a magic square in the sense that the sum of the numbers in each row and each column is exactly the same, namely, 260, yet it is not quite a magic square because the two main diagonals fail to have this same sum. Such a square is called *semi-magic*. Also—and this is a very minor defect indeed—Euler has used an open knight's tour, rather than a knight's tour. It should be pointed out, however, that Euler's semi-magic square has other admirable qualities that more than make up for these two deficiencies: note that the four quadrants of Euler's 8×8 semi-magic square are themselves 4×4 semi-magic squares; and that, in turn, each 2×2 quadrant within the 4×4 quadrants contains four numbers that add up to the same sum, namely, 130.

In Figure 4.5 we show two 8×8 semi-magic squares that use knight's tours for their construction, one by Jaenisch in 1862 and the other by Wenzelides. Note that both of these knight's tours use an idea we described in Problem 2.2 in Chapter 2 and

1	48	31	50	33	16	63	18
30	51	46	3	62	19	14	35
47	2	49	32	15	34	17	64
52	29	4	45	20	61	36	13
5	44	25	56	9	40	21	60
28	53	8	41	24	57	12	37
43	6	55	26	39	10	59	22
54	27	42	7	58	23	38	11

Figure 4.4 An 8×8 open knight's tour semi-magic square by Euler.

46	55	44	19	58	9	22	7
43	18	47	56	21	6	59	10
54	45	20	41	12	57	8	23
17	42	53	48	5	24	11	60
52	3	32	13	40	61	34	25
31	16	49	4	33	28	37	62
2	51	14	29	64	39	26	35
15	30	1	50	27	36	63	38

50	11	24	63	14	37	26	35
23	62	51	12	25	34	15	38
10	49	64	21	40	13	36	27
61	22	9	52	33	28	39	16
48	7	60	1	20	41	54	29
59	4	45	8	53	32	17	42
6	47	2	57	44	19	30	55
3	58	5	46	31	56	43	18

Figure 4.5 8×8 knight's tour semi-magic squares by Jaenisch, left, and Wenzelides, right.

connect the individual 4-cycles in each of the 4×4 quadrants to form the knight's tour; this has the salubrious effect of evenly distributing numbers in the rows and columns.

Jaenisch, however, did manage to construct an 8×8 magic square using two knight's cycles, each covering exactly half of the chessboard, another near miss! In Figure 4.6 we can see that he first places the numbers 1 through 32 in a single cycle using knight's moves. Then he places the numbers 33 though 64 in a completely separate cycle in the remaining 32 squares.

And yet, the big question has remained open: is there is an 8×8 knight's tour magic square? It is known that any square board with a magic knight's tour would have to have sides that are a multiple of 4, so it is also an open question as to whether there is a 12×12 knight's tour magic square? Astonishingly,

15	20	17	36	13	64	61	34
18	37	14	21	60	35	12	63
25	16	19	44	5	62	33	56
38	45	26	59	22	55	4	11
27	24	39	6	43	10	57	54
40	49	46	23	58	3	32	9
47	28	51	42	7	30	53	2
50	41	48	29	52	1	8	31

Figure 4.6 An 8×8 magic square by Jaenisch using two 32-cycle knight's tours.

however, knight's tour magic squares have been constructed on chessboards of size 16×16, 20×20, 24×24, 32×32, 48×48, and 64×64. A 16×16 knight's tour magic square is shown in Figure 4.7. However, it has just been announced by Guenter Stertenbrink that a team of computers, exhaustively searching all possibilities and using a program written by J. C. Meyrignac, concluded on 5 August 2003 that no such knight's tour exists for the 8×8 chessboard. Details of this announcement can be found at http://magictour.free.fr.

Using Kings and Rooks to Construct Magic Squares

In 1921 Ghersi constructed the 8×8 magic square shown in Figure 4.8 using a king's tour of the chessboard. First, note the pattern of the numbers 1–8 in his tour. He then exactly reversed this pattern for the numbers 9–16 in a symmetric fashion in the second quadrant. He then placed 17–32 in the lower half of the board as a mirror image of 1–16, again reversing the order. Finally, he placed 33–64 as a mirror image, left to right, of 1–32, also in reverse order.

Problem 4.4 Construct an 8×8 semi-magic square using a king's tour starting with the numbers 1–8 placed in the 2×4 rectangular block in the upper right-hand corner of an 8×8 chessboard. In other words, replace the 54 in Figure 4.8 in Ghersi's tour with a 6.

184	217	170	75	188	219	172	77	228	37	86	21	230	39	88	25
169	74	185	218	171	76	189	220	85	20	229	38	87	24	231	40
216	183	68	167	222	187	78	173	36	227	22	83	42	237	26	89
73	168	215	186	67	174	221	190	19	84	35	238	23	90	41	232
182	213	166	69	178	223	176	79	226	33	82	31	236	43	92	27
165	72	179	214	175	66	191	224	81	18	239	34	91	30	233	44
212	181	70	163	210	177	80	161	48	225	32	95	46	235	28	93
71	164	211	180	65	162	209	192	17	96	47	240	29	94	45	234
202	13	126	61	208	15	128	49	160	241	130	97	148	243	132	103
125	60	203	14	127	64	193	16	129	112	145	242	131	102	149	244
12	201	62	123	2	207	50	113	256	159	98	143	246	147	104	133
59	124	11	204	63	114	1	194	111	144	255	146	101	134	245	150
200	9	122	55	206	3	116	51	158	253	142	99	154	247	136	105
121	58	205	10	115	54	195	4	141	110	155	254	135	100	151	248
8	199	56	119	6	197	52	117	252	157	108	139	250	153	106	137
57	120	7	198	53	118	5	196	109	140	251	156	107	138	249	152

Figure 4.7 A 16×16 knight's tour magic square.

61	62	63	64	1	2	3	4
60	11	58	57	8	7	54	5
12	59	10	9	56	55	6	53
13	14	15	16	49	50	51	52
20	19	18	17	48	47	46	45
21	38	23	24	41	42	27	44
37	22	39	40	25	26	43	28
36	35	34	33	32	31	30	29

Figure 4.8 Ghersi's 8×8 king's tour magic square.

In 1985 Stanley Rabinowitz found the 8×8 magic square shown in Figure 4.9 using a rook's tour of the chessboard. This really is a rook's tour. In Chapter 1 the rook's tours we used to prove Gomory's Theorem were of a special kind where the rook moved a single square at a time. But, for example, in Fig-

61	62	63	64	1	2	3	4
12	11	10	9	56	55	54	53
20	19	18	48	17	47	46	45
60	59	58	8	57	7	6	5
37	38	39	25	40	26	27	28
13	14	15	49	16	50	51	52
21	22	23	24	41	42	43	44
36	35	34	33	32	31	30	29

Figure 4.9 Rabinowitz's 8×8 rook's tour magic square.

ure 4.9, the move from 4 to 5 is a legal rook's move. Note that the pattern of the numbers 1–8 is repeated for the numbers 9–16 in almost a mirror image, left to right, but shifted down. The rest of the construction is just like Ghersi's method above: that is, 17–32 are placed as a mirror image, top to bottom, of 1–16, reversing the order; and 33–64 are placed as a mirror image, left to right, of 1–32, in reverse order.

Solutions to Problems

Solution 4.1 The two magic squares are shown in Figure 4.10.

11	24	7	20	3
4	12	25	8	16
17	5	13	21	9
10	18	1	14	22
23	6	19	2	15

22	47	16	41	10	35	4
5	23	48	17	42	11	29
30	6	24	49	18	36	12
13	31	7	25	43	19	37
38	14	32	1	26	44	20
21	39	8	33	2	27	45
46	15	40	9	34	3	28

Figure 4.10 Muhammad ibn Muhammad's 5×5 and 7×7 magic squares.

Solution 4.2 Muhammad ibn Muhammad's 7×7 magic square using the knight's move is shown in Figure 4.11.

32	14	38	20	44	26	1
48	23	5	29	11	42	17
8	39	21	45	27	2	33
24	6	30	12	36	18	49
40	15	46	28	3	34	9
7	31	13	37	19	43	25
16	47	22	4	35	10	41

Figure 4.11 Muhammad ibn Muhammad's 7×7 magic square using the knight's move.

Solution 4.3 The knight's tour and the resulting magic square are shown in Figure 4.12.

Solution 4.4 The desired king's tour semi-magic square is shown in Figure 4.13.

1	116	49	151	84	17	119	65	167	100	33	135	68
136	69	2	117	50	152	85	18	120	53	168	101	34
102	35	137	70	3	105	51	153	86	19	121	54	169
55	157	103	36	138	71	4	106	52	154	87	20	122
21	123	56	158	104	37	139	72	5	107	40	155	88
156	89	22	124	57	159	92	38	140	73	6	108	41
109	42	144	90	23	125	58	160	93	39	141	74	7
75	8	110	43	145	91	24	126	59	161	94	27	142
28	143	76	9	111	44	146	79	25	127	60	162	95
163	96	29	131	77	10	112	45	147	80	26	128	61
129	62	164	97	30	132	78	11	113	46	148	81	14
82	15	130	63	165	98	31	133	66	12	114	47	149
48	150	83	16	118	64	166	99	32	134	67	13	115

Figure 4.12 A 13×13 knight's tour magic square.

61	62	63	64	1	2	3	4
60	59	58	57	8	7	6	5
12	11	10	9	56	55	54	53
13	14	15	16	49	50	51	52
20	19	18	17	48	47	46	45
21	22	23	24	41	42	43	44
37	38	39	40	25	26	27	28
36	35	34	33	32	31	30	29

Figure 4.13 An 8×8 king's tour semi-magic square.

CHAPTER FIVE

The Torus and the Cylinder

In this chapter we will look at a number of the many variations on the theme of knight's tours that have been explored by considering chessboards on two surfaces: the torus and the cylinder. We begin with the most natural variation of all, one which we have discussed several times already, and which involves simply wrapping both the columns and the rows of the chessboard into closed loops.

Chessboards on the Torus

A *torus* is the name we give to a surface that is shaped like a doughnut. We can think of creating a torus by starting with a rectangle, say a piece of paper, and first folding the top and bottom edges of this rectangle together to form a cylindrical tube, and then bending and stretching this tube so that the left and right ends of the tube also come together; in other words, in addition to folding the top and bottom edges of the rectangle together, we have also folded the left and right edges of the rectangle together.

A *torus*—or, more precisely, a *flat torus*—is, thus, the name we give to this rectangle with the top and bottom edges identified, and with also the left and right edges identified. Almost always, we carry out these identifications only in our imaginations, that is, the rectangle stays flat. And we usually blur the distinction between what are in fact two quite different objects, the torus, which is an actual three-dimensional surface; and the flat torus, which is a flat two-dimensional rectangle with sides identified, and we frequently refer to either one of these objects

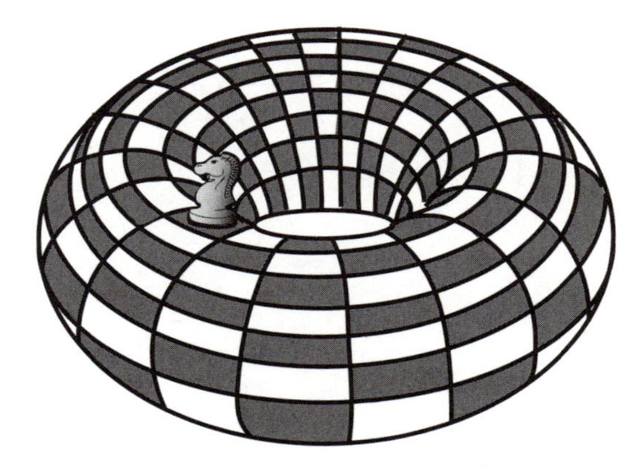

Figure 5.1 A toroidal chessboard.

as simply *the torus*. For example, we might speak of playing chess 'on a torus' and perhaps might even occasionally visualize, say, a knight touring a real three-dimensional toroidal chessboard as shown in Figure 5.1, but more often than not we will be thinking in terms of a perfectly flat two-dimensional chessboard that just happens to have special rules for movement of the chess pieces: that is, we are thinking of a flat torus.

Problem 5.1 Since tic-tac-toe is played on a 3×3 rectangular board, we can modify the normal rules and play this game on a torus instead. The object will still be to get 'three in a row', but now, on a torus, there are more ways to get 'three in a row' since there are additional diagonal winning positions. Show that if tic-tac-toe is played on a torus, the first player can always win the game no matter how the second player responds.

Problem 5.2 It is obviously never possible on a regular chessboard for a bishop and a knight to simultaneously threaten one another simply because a knight has an '*L*'-shaped move and a bishop moves along diagonal 45° lines. Is it possible for a bishop and a knight to simultaneously threaten one another on a toroidal chessboard?

There is an excellent website where you can play online games such as tic-tac-toe and chess on a variety of surfaces, including the torus and a surface we haven't yet discussed, the Klein bottle: www.northnet.org/weeks/SoS. This site is a

1	2	3	4	5	6	7
9	8	14	13	12	11	10

1	2	3	4	5	6

1	16	7	22	13	4	19	10
20	11	2	17	8	23	14	5
15	6	21	12	3	18	9	24

1	11	3	13	5	15	7	17	9
29	19	27	35	25	33	23	31	21
10	2	12	4	14	6	16	8	18
20	28	36	26	34	24	32	22	30

Figure 5.2 Knight's tours on a torus for 1×6, 2×7, 3×8, and 4×9 chessboards.

companion to a wonderful book on surfaces and space by Jeffrey R. Weeks called *The Shape of Space*, which itself is well worth looking at if you don't already know of it.

We previously encountered toroidal chess in Chapter 4 when we discussed the work of Muhammad ibn Muhammad. Later in the book we will see that G. Pólya used a similar idea almost 100 years ago while working on the 'n-queens problem', but let us return for the moment to the Knight's Tour Problem. I did mention in Chapter 1 that there is a knight's tour for a 5×5 chessboard on a torus, as we could see in Figure 1.7. This is not really too surprising: with the additional mobility that a knight gains on a torus, chessboards that don't ordinarily have knight's tours might indeed have knight's tours if we place the board on a torus. It is a bit surprising, however, exactly how beneficial this extra mobility turns out to be. In fact, what is true is the following.

Theorem 5.1 (Watkins and Hoenigman) *On a torus, every rectangular chessboard has a knight's tour.*

This remarkable result was proved [33] by a student, Becky Hoenigman, and me in a manner very similar to that used by Schwenk to prove Theorem 3.1. We first showed that any $1 \times n$, $2 \times n$, $3 \times n$, or $4 \times n$ chessboard could be toured on a torus, and then showed how to stack these tours together to build tours for chessboards of arbitrary size. Knight's tours for four of these smaller boards are shown in Figure 5.2. In Problem 5.3 you can try your hand at stacking several of these tours together to form larger tours.

Problem 5.3 Find a knight's tour on a torus for a 6×8 chessboard, and also for an 8×9 chessboard.

GRAY CODES AND THE 4-CUBE

Before we turn to look at knight's tours on still more surfaces, let's take one more look at the 4×4 chessboard, this time on a torus. As we have previously observed, a knight has significantly more mobility on a torus, and so the knights graph for the 4×4 chessboard on a torus is going to be considerably richer than the knights graph shown either in Figure 1.6 in Chapter 1 or in Figure 3.2 in Chapter 3. In particular, on a torus each square is now completely equivalent, and so the degree of each vertex will be 4. Moreover, and this is something you will be asked to verify in Problem 5.4, it turns out that the 4×4 toroidal knights graph is identical to an extremely famous graph called the 4-cube.

The 4-*cube* is the graph that represents a four-dimensional cube and it is shown in Figure 5.3 along with the 1-cube, the 2-cube, and the 3-cube. The first three of these graphs, rather naturally, correspond to a line segment, a square, and a cube. By the way, a four-dimensional cube is also sometimes called a hypercube. Note how in going from any one dimension in these graphs to the next higher dimension, we can think of creating the next higher order 'cube' from the existing 'cube' by adding a 0 in front of each vertex of the existing 'cube' to create one set of vertices for the new 'cube', and then moving the existing 'cube' in an entirely new direction, or dimension, and adding a 1 in front of each vertex of the existing 'cube' in this new position to create a second set of vertices for the new 'cube'. Similarly, the new 'cube' has edges from the existing 'cube' in both the old and new positions as well as edges created by the process of moving each vertex from its old to its new position, which also creates a new edge—think of the motion as causing the vertex to trace out a new edge. In this way, in turn, the square is created from the line, the cube is created from the square, the four-dimensional cube from the cube, the five-dimensional cube from the four-dimensional cube, and so on.

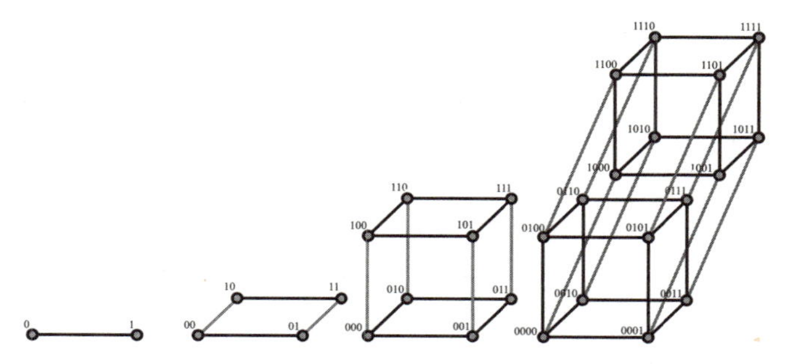

Figure 5.3 The first four cube-graphs: the 1-cube, 2-cube, 3-cube, and 4-cube.

Gray codes were invented by Frank Gray in the 1940s in the course of his work on electro-optical systems for Bell Telephone Laboratories. The idea is to find a way to transmit strings of data so that a small error occurring during transmission will result in only a small error in the received message. Gray's ingenious way of doing this was to list all the strings in a sequence so that adjacent strings differ in only one position. For example, if we take all eight binary strings of length 3, we can list them in the following Gray code order:

$$000–001–011–010–110–111–101–100.$$

Note that this list is actually a cycle since the last entry, 100, can also comes back to the very first entry, 000, since these two entries also differ in a single position. It should also be clear to you that this particular Gray code in fact corresponds to a Hamiltonian cycle in the 3-cube in Figure 5.3. This Gray code is called the *standard binary reflected Gray code* because it first traverses the bottom square face of the cube, and then shifts to the top square face, which it then traverses in reverse order, before shifting back to the bottom in order to close the cycle. In fact, this code did the exact same thing for the bottom square itself, doing the front edge first, shifting to the back edge, then doing the back edge in reverse order, and finally shifting back to the front in order to close the tour.

Problem 5.4 First, find a Gray code for the sixteen binary strings of length 4. Then, by labeling the squares of a 4×4

Figure 5.4 White to mate in four moves on a torus.

chessboard so that the knights graph for the 4×4 chessboard on a torus becomes the 4-cube, show that your Gray code also provides a knight's tour on a torus for the 4×4 chessboard.

A Chess Problem

I promised in the very first chapter that this book was to be about mathematics, and not about chess, but I can't resist including the following chess puzzle taken from a collection of puzzles by Miodrag Petković, *Mathematics and Chess* [24]. This lovely puzzle has the standard form for chess problems so well-known to chess enthusiasts around the world: white to play and mate in four moves; the only difference here is that the chessboard is now on a torus—but what a difference!

Problem 5.5 Given the position of the chess pieces on the toroidal chessboard in Figure 5.4, in which it is White's turn to play, find a way for White to win in just four moves, no matter what Black's responses are.

Chessboards on Cylinders

You might have noticed that several of the toroidal knight's tours we have previously come across did not in fact make full use of both the horizontal and the vertical wrapping of the chessboard. For example, in the tour of the 5×5 board on a torus indicated in Figure 1.7 in Chapter 1, the knight used only the vertical wrapping of the columns to get back to square 1. Similarly, the knight's tour of the 3×8 board on a torus shown in Figure 5.2 in this chapter made use of only the horizontal wrapping of the rows. In other words, each of these particular two knight's tours would work just as well on a cylindrical chessboard, that is, a board in which just one pair of opposite edges have been identified. The idea of cylindrical chess may be very old, certainly well over 100 years, and a number of writers have discussed knight's tours on cylinders and have provided tours for several of the cylindrical boards that cannot be achieved on regular chessboards of the same size.

Problem 5.6 See's Candies uses several other chessboard patterns in addition to those mentioned in Problem 3.2. For example, they have a 3×100 cylindrical chessboard pattern on their one pound can of See's Toffee-Ettes. Find a knight's tour for this cylindrical chessboard.

So, at this point, we might even begin to hope that all chessboards could be toured on a cylinder. That is not the case, however, and an exact description of which rectangular chessboards have knight's tours on cylinders is given by the following theorem.

Theorem 5.2 (Watkins) *An $m \times n$ cylindrical chessboard with m rows and n columns—the rows wrapped around the cylinder—has a knight's tour unless one of the following two conditions holds:*

(a) $m = 1$ *and* $n > 1$*; or*

(b) $m = 2$ *or* 4 *and* n *is even.*

It is easy to see why these cases are impossible on a cylinder. If $m = 1$, a knight can't move at all. If $m = 2$, each move takes a knight either right or left by two columns, and so, if n is even, a knight could visit at most half of the columns. If $m = 4$ and n is even, then the coloring arguments of Pósa and Golomb given in Chapter 3 still work here on cylinders to show that a knight's tour is impossible because their clever colorings successfully wrap around the cylinder precisely because n is even.

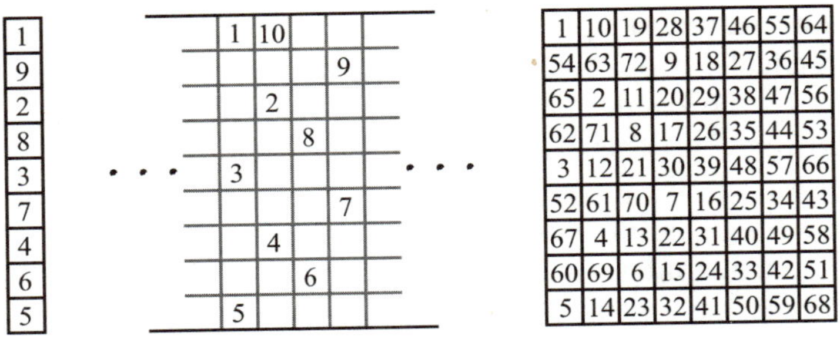

1
9
2
8
3
7
4
6
5

. . .

	1	10			
				9	
		2			
			8		
	3				
				7	
		4			
			6		
	5				

. . .

1	10	19	28	37	46	55	64
54	63	72	9	18	27	36	45
65	2	11	20	29	38	47	56
62	71	8	17	26	35	44	53
3	12	21	30	39	48	57	66
52	61	70	7	16	25	34	43
67	4	13	22	31	40	49	58
60	69	6	15	24	33	42	51
5	14	23	32	41	50	59	68

Figure 5.5 A simple repeating pattern that tours an $m \times n$ cylindrical chessboard when m is odd.

The proof of Theorem 5.2 given in [32] showed that all chessboards other than those excluded above have knight's tours by producing tours for any chessboards remaining from each of the three cases of Schwenk's Theorem 3.1. A somewhat nicer argument is given by Ian Stewart in [28] in which he splices together easy repeating patterns for $3 \times n$ and $5 \times n$ boards. This leaves only boards of sizes $2 \times n$, $4 \times n$, and $7 \times n$ for him to deal with separately.

In fact, it turns out to be quite easy to generalize the idea behind Stewart's repeating patterns for $3 \times n$ and $5 \times n$ boards and to find repeating patterns for any $m \times n$ board where m is odd. We illustrate the idea with a tour for a $9 \times n$ board. Begin on the left in Figure 5.5 with a tour for a 9×1 board using a pattern that works for any $m \times 1$ board where m is odd. This tour, by the way, is essentially unique, the only other option being to go in the reverse direction. Note that this tour mostly involves knight's moves that take two steps up or down and only one step right or left. In fact, only two of the knight's moves take two horizontal steps, and the rest take one horizontal step. So, now, it is easy to create a repeating pattern on the $9 \times n$ chessboard. Begin with 1 somewhere in the top row. There are two knight's moves with two horizontal steps for us to do somewhere; do one of these to the left and one of these to the right. There are an odd number of knight's moves with one horizontal step for us to do; be sure to do one more of these to the right than to the left. One of many ways of doing all of this is shown in Figure 5.5. Now, starting at 10, you can just repeat this pattern over and over again to finish a tour of the board. The resulting tour for a 9×8 board is also shown in Figure 5.5.

Unfortunately, the same trick doesn't quite work if m is even. You can still tour the $m \times 1$ board, but now there are an even number of knight's moves with one horizontal step, so you can't make a shift of exactly one square to the right work for a repeating pattern. This isn't a major obstacle, however, since if m is even you can just splice a tour of a $3 \times n$ board on top of a tour of an $(m - 3) \times n$ board.

Finally, I should also mention that touring a $2 \times n$ board where n is odd is automatic, just wind the knight up and turn him loose. And for the $4 \times n$ board where n is odd, all you have to do is splice two $2 \times n$ tours together.

SOLUTIONS TO PROBLEMS

Solution 5.1 Since on a torus all squares are equivalent, the first player might as well place an X in the center square of the board. By symmetry, the second player then has exactly two options, either place an O horizontally right next to the X or diagonally right next to the X. Either way, the first player can then force a win as shown in Figure 5.6. In each case, the second player is forced on the second move to block a winning threat, and then, in the final position in each case, the second player is unable to simultaneously block the two different winning threats, and therefore loses on the next move.

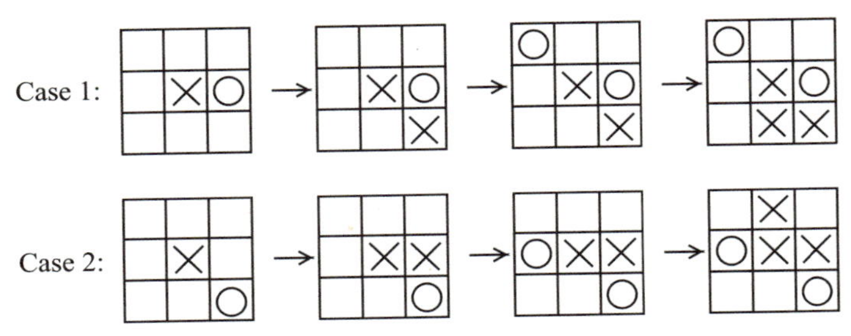

Figure 5.6 The first player wins.

Solution 5.2 Yes, it is possible. For example, on a 3×3 chessboard a bishop and a knight placed in opposite corners threaten one another. In general, however, the answer depends on the size of the chessboard. For instance, for an 8×8 board, the answer is: no, even on a torus, it is still not possible for a bishop and a knight to simultaneously threaten one another. This is easy to prove using the standard black and white checkerboard coloring of the 8×8 chessboard.

A bishop is always restricted to squares of a single color and a knight on a square of one color only threatens squares of the other. This remains true even when they go off the board on one side and reappear on another. This argument using coloring can obviously be applied for any board where both dimensions of the board are even because having an even number of squares in a row or column ensures that the color pattern is preserved when the row or column is wrapped around on itself.

1	16	7	22	13	4	19	10
20	11	2	17	8	23	14	5
15	6	21	12	3	18	9	24

1	16	7	22	13	4	19	10
20	11	2	17	8	23	14	5
15	6	21	12	3	18	9	24

1	11	3	13	5	15	7	17	9
29	19	27	35	25	33	23	31	21
10	2	12	4	14	6	16	8	18
20	28	36	26	34	24	32	22	30

1	11	3	13	5	15	7	17	9
29	19	27	35	25	33	23	31	21
10	2	12	4	14	6	16	8	18
20	28	36	26	34	24	32	22	30

Figure 5.7 Stacking tours on a torus.

Solution 5.3 This can be done by stacking tours from Figure 5.2, using two copies of the 3×8 board and two copies of the 4×9 board as shown in Figure 5.7. On the left we have two identical copies of the tour for the 3×8 board in which the move 14–15 has been highlighted in the top board, and the move 18–19 highlighted in the bottom board. We can tour the entire 6×8 board by beginning at 1 in the top board, touring that board until we reach 14 and then jumping to 18 on the bottom board, which after all is a completely legitimate move for a knight on a torus! From there, we tour the entire bottom board in reverse all the way from 18 until we get to 19, and then from 19 we can jump back to 15 on the top board and finish completing the tour there.

On the right in Figure 5.7 we have two identical copies of the tour for the 4×9 board with the move 1–36 highlighted in each board. We can tour the entire 8×9 board by beginning at 1 in the top board, completely touring that board ending at 36, and then jumping to 1 on the bottom board, completely touring the bottom board ending at 36 in the bottom row, from where we can complete the tour by jumping back to 1 at the very top.

Solution 5.4 Start with a Gray code for the binary strings of length 3, namely, 000–001–011–010–110–111–101–100, and form a Gray code for the sixteen binary strings of length 4 by first placing a 0 in front of each string in the Gray code for the eight strings of length 3, and then follow this immediately,

0000	0111	1001	1110
1011	1100	0010	0101
0110	0001	1111	1000
1101	1010	0100	0011

1	6	15	12
14	9	4	7
5	2	11	16
10	13	8	3

Figure 5.8 A 4-cube labeling of a toroidal chessboard and the resulting knight's tour.

but in reverse order, by placing a 1 in front of each string of length 3. The result is

0000-0001-0011-0010-0110-0111-0101-0100-
1100-1101-1111-1110-1010-1011-1001-1000,

which is the desired Gray code. Then, in Figure 5.8 we label the 4×4 board with the sixteen binary strings of length 4 as shown on the left. Then, by going in the Gray code order above we get the toroidal knight's tour shown on the right in Figure 5.8.

Solution 5.5 Since this is a chess puzzle, after all, I will use the standard chess notation which labels the rows 1–8 from the bottom to the top and labels the columns a–h from the left to the right. Thus, a1 is the square in the lower left-hand corner and h1 is the square in the lower right-hand corner. White's first play is to move the white queen to h7. Note that this traps the black queen along the top row since the presence of the white king is preventing the black king from moving up. Thus, the black king has only two choices, move left or right one square.

If the black king moves to the right, that is, to f8, the white queen backs up a step to g6. The black king has only one move, to e7. The white king moves down a row to e1, further squeezing the black king, who is forced to d7. The white queen moves to e8 for the checkmate.

If, on the other hand, the black king originally moves to the left, that is, to d8, the white king moves across to c7. This places the black king in check and forces him back to e8. The white knight moves to h6. The black king, still trapped along the top row, must move to f8. The white queen moves to e1 for the checkmate.

Solution 5.6 Here in Figure 5.9 are two simple repeating patterns that tour the See's Candies 3×100 cylindrical chessboard. There are several others.

	298	1	4	7	10	13	16	
. . .	293	296	299	2	5	8	11	. . .
	300	3	6	9	12	15	18	

	298	199	100	1	202	103	4	205	106	7	208	109	10	
. . .	197	98	299	200	101	2	203	104	5	206	107	8	209	. . .
	96	297	198	99	300	201	102	3	204	105	6	207	108	

Figure 5.9 Two cylindrical knight's tours.

CHAPTER SIX

The Klein Bottle and Other Variations

Up to this point we have looked at rectangular chessboards in three different ways. The first way is as the perfectly ordinary chessboard with standard edges that fully contain the chess pieces within the chessboard. The second way is to identify the top and the bottom edges of the board and also to identify the left and the right edges, that is, the chessboard becomes a flat torus and the chess pieces gain considerable freedom of movement since the edges, in effect, disappear. The third way to look at rectangular chessboards is to identify just one pair of opposite edges, in which case we have a cylindrical board.

THE KLEIN BOTTLE

I think it might occur to you that there may be other interesting ways to identify edges of a rectangular chessboard as well. There are. Here is a fourth way: identify the top and bottom edges of the rectangle just as we did before for the torus; and also identify the left and the right edges, but this time in reverse order! So the top and bottom are identified in the normal way, but the sides get a half-twist before they are identified. A rook in the left-most column moving up and going off the top of the board therefore reappears at the bottom of the board in the same column; but a rook in the bottom row going off the right side of the board will reappear at the upper left in the top row!

This surface—and it is a surface—is called the *Klein bottle*. The Klein bottle exists, but unfortunately not in the three-dimensional space we are most familiar with. You can see the difficulty if you imagine trying to actually construct a Klein bottle by first folding a rectangle into a cylinder, which is easy.

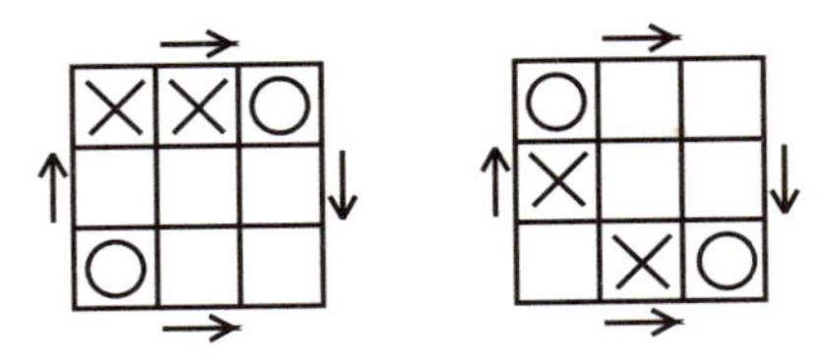

Figure 6.1 What is X's best play in each of these games of Klein bottle tic-tac-toe?

Next, you need to bend this cylinder as we did before for the torus and identify the ends, but this time in reverse order, which means that one end needs to pass inside the cylinder first and attach to the other end from the opposite direction. In your mind's eye the result should look kind of bottle shaped. Fortunately, instead, we can work quite easily with the flat rectangular version of the Klein bottle, although we often add a few arrows to our diagrams to remind ourselves of the peculiar identification of edges that is taking place.

In order to get a feeling for this surface, let's analyze X's position in each of the games of tic-tac-toe in Figure 6.1. In the first game you might think X should block O by playing in the middle square. But, in fact, X has an immediate winning play instead! Simply follow the horizontal line in the top row through the two squares that each contain an X as this line goes off the left side of the board. Where does this line reappear? Because of the twist in the Klein bottle, the line reappears in the lower right corner, so placing an X in the lower right corner makes 'three in a row' for X. This is shown on the left in Figure 6.2.

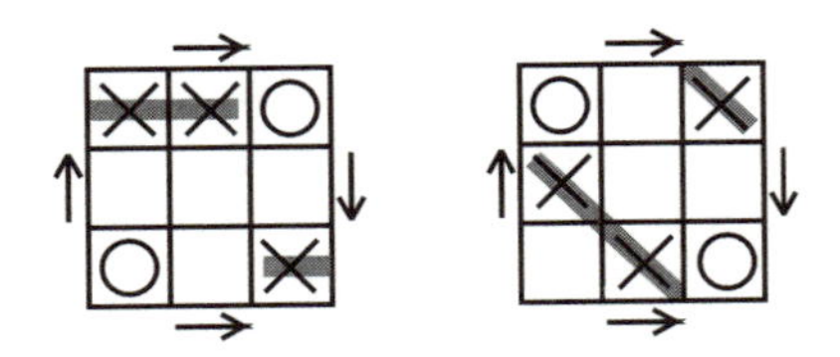

Figure 6.2 X's winning plays.

In the second game in Figure 6.1, X again has an immediate winning play. The diagonal line going up and left through the two squares each containing an X is blocked by the O in the

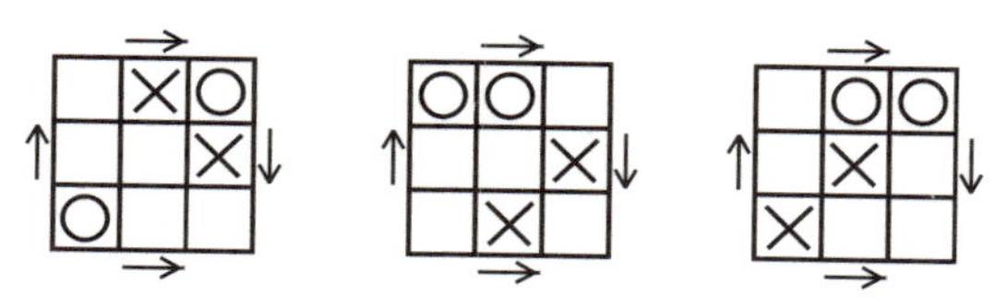

Figure 6.3 What is X's best play?

lower right-hand corner. But the same diagonal line going down and to the right through the two squares each containing an X reappears at the top of the board in the upper right-hand corner. This 'three in a row' is shown on the right in Figure 6.2.

Problem 6.1 Find X's best play in each of the games of Klein bottle tic-tac-toe shown in Figure 6.3.

If you care to, you can get more practice playing with the Klein bottle online at www.northnet.org/weeks/SoS. In any event, you should now be warmed up enough on this strange surface to prove the following theorem.

Theorem 6.1 (Watkins) *On a Klein bottle, every rectangular chessboard has a knight's tour.*

Because of Theorem 5.2, we need only provide knight's tours on a Klein bottle for the instances of rectangular boards where a knight's tour does not already exist on a cylinder. Each of these instances is illustrated in Figure 6.4 and the patterns indicated by these specific examples are completely general, although note that when $n = 2$ or 4 and m is even the pattern does depend on whether m is congruent to 0 or 2 modulo 4. Note also that since for a Klein bottle we are now wrapping the columns around the cylinder, the roles of m and n are reversed from their roles in Theorem 5.2. Following a knight along these five tours is a pretty good way to make sure you understand the Klein bottle. In particular, make sure you understand move 6–7 on the 6×2 board; it is the same move as 16–17 on the 8×4 board.

THE MÖBIUS STRIP

The next way for us to look at a rectangular chessboard—this is the fifth, and last, if you're counting—is to again identify just

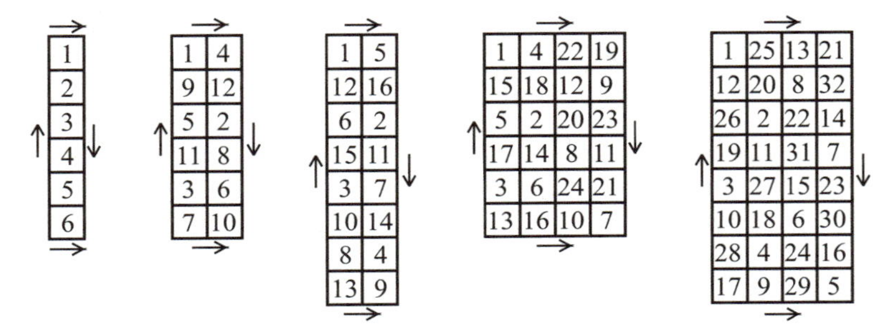

Figure 6.4 Knight's tours on a Klein bottle.

one pair of opposite edges as we did for the cylindrical board, but this time to add a half twist just as we did for the Klein bottle. The resulting surface is known as the Möbius strip. And, of course, there is a theorem to be told, one that is analogous to Theorem 5.2.

Theorem 6.2 (Watkins) *An $m \times n$ chessboard on a Möbius strip with m rows and n columns—the rows wrapped around the Möbius strip—has a knight's tour unless one or more of the following three conditions hold:*

(a) $m = 1$ *and* $n > 1$*; or* $n = 1$ *and* $m = 3, 4,$ *or* 5*;*

(b) $m = 2$ *and* n *is even, or* $m = 4$ *and* n *is odd;*

(c) $n = 4$ *and* $m = 3$.

The Möbius strip is the most famous example of what is called a one-sided surface, because, as M. C. Escher has made so clear for us, an ant walking once around the Möbius strip appears to come back on the 'other' side; in other words, there is really only one side to this surface. This raises an issue for chess pieces. If a chess piece, like Escher's ant, travels once around a Möbius strip, or a Klein bottle for that matter, and lands back on the original square, but on the 'other' side, is this the same square or a new square? Well, it's the same square, the confusion arises only because of our casual use of the word 'on'. Strictly speaking, since the Möbius strip is a two-dimensional surface, you should be thinking of chess pieces as two-dimensional objects moving inside of the surface, and so it is more appropriate to say a piece is 'at' or 'in' a given square, rather than 'on' it.

Euler's Formula

You may have been wondering all along about our particular choice of surfaces on which to place chessboards. Why not put a chessboard on a sphere, for example? You may have also noticed a significant difference between, say, chessboards on a torus or a Klein bottle and those placed on a cylinder or a Möbius strip or ordinary chessboards. The latter three boards all have boundaries which the chess pieces are not allowed to cross, whereas on the torus or the Klein bottle the boards no longer have any boundaries or edges at all. This puts the torus and the Klein bottle into very select company among surfaces—along with the sphere and the infinite plane—as being of particular interest to mathematicians because locally they all look the same at any given point on the surface. From the outside, these surfaces look quite different, but if you zoom in really close to one of these surfaces it would be very hard to tell which surface you were actually on. So, from a mathematician's point of view, the torus and the Klein bottle are both extremely important surfaces to study. But, let me ask again, why not put a chessboard on a sphere? It's an important surface, too. The answer can be found in Euler's formula.

Since about the middle of the nineteenth century mathematicians have had a complete list of all possible surfaces. This list is fully described by what is known as *The Classification Theorem* and the list tells us precisely what surfaces can—and can't—look like. In particular, it tells us what *closed* surfaces can be like. Surfaces such as the sphere, the torus, and the Klein bottle are called *closed* because they close in on themselves, meaning, essentially, that they are finite. This is in contrast to a surface such as the plane which is infinite or open. These closed surfaces in which we are interested are usually more formally called *two-manifolds*.

First of all, then, The Classification Theorem tells us that there are two kinds of two-manifolds, those that are orientable and those that are non-orientable. Surfaces such as the sphere and the torus are called orientable because we can assign an orientation to them—for example, 'clockwise'—that is reliable throughout the surface. If we take a clock on a trip around the world—or around a torus, for that matter—when we get back home it will still be running in the same direction. But if we take a clock on a trip once around a Möbius strip it will be running in the opposite direction when we get back. The same bizarre thing can happen on other non-orientable surfaces such as the Klein bottle and the projective plane. The *projective plane* is a surface that is formed by taking a hemisphere and identifying all pairs of opposite points on the rim of the hemisphere.

With this bit of background, the classification of surfaces is now remarkably simple and beautiful. The orientable part of the list of surfaces is based upon the torus and consists of the sphere (built from zero tori), the torus (built from one torus), the two-holed torus (built from two tori—think of a doughnut with two holes), the three-holed torus (built from three tori), and so on. That's it: an infinite list of surfaces built from more and more tori. The non-orientable part of the list of surfaces is the same story: the projective plane (built from one projective plane), the Klein bottle (built from two projective planes), and so on. That's it: another infinite list built from more and more projective planes. The Classification Theorem says if you have a two-manifold, then it is one of the surfaces on one of these two lists. There are no others! By the way, it is worth noting that the cylinder and the Möbius strip are surfaces with boundaries and so, strictly speaking, they don't fall within this classification scheme, but they of course fit into the larger scheme in a very natural way.

Now, at last, we can discuss Euler's formula. It turns out that each of these surfaces, or two-manifolds, has a number that can be assigned to it, called its *Euler characteristic*. And these numbers are distinct! Well, at least every orientable surface has a distinct number, and every non-orientable surface has a distinct number. This is exactly like numbers on athletic jerseys. If during a football game I were to say 'number 7 for the Denver Broncos', you would know exactly who I was talking about, even though the opposing team may also have a 'number 7'. In other words, the Euler characteristic is a number, or a label, that these surfaces carry around on themselves so that we can tell them apart. Isn't that convenient? The numbers for the orientable surfaces are 2, 0, −2, −4, −6, ..., and the numbers for the non-orientable surfaces are 1, 0, −1, −2, −3, ... Note that some of these numbers are unique. If I say to you 'surface number −5', you know exactly which surface I am talking about. But if I say 'surface number −2', you don't know which of two surfaces I might mean: the one on the orientable team or the one on the non-orientable team. I would need to tell you.

Euler found a formula that computes or 'reads' these numbers and, therefore, he found a way for us to tell these surfaces apart. Here is how it works. Suppose you have a surface and you want to know its Euler characteristic. First you draw a graph on your surface without letting any edges cross and with the following property: every region formed by the graph must look like a polygonal 'disk'. In other words, you could cut each region out of the surface along edges of the graph and then flatten each region out in the plane and it would look like a polygon. Then, Euler said all you have to do is count the number of vertices, v, in your graph, the number of edges, e, and the number of regions, r, and then the Euler characteristic is given by *Euler's formula*:

$$\text{the Euler characteristic of the surface} = v - e + r.$$

So, what can this formula tell us about chessboards? When I decide to put a chessboard on one of these surfaces—and remember, these surfaces don't have boundaries—I want the following thing to happen: I want every square on the board to 'feel' like a square in the middle of a standard chessboard. In other words, chess pieces should be able to move in any one of the four 'horizontal/vertical' directions and in any one of the four 'diagonal' directions. It doesn't bother me that the square to my right may actually turn out to be the same square as the one to my left, the point is that I can move to my right if I want, just like I could if I were in the middle of a standard chessboard.

Now, a chessboard on one of these surfaces can itself be thought of as a graph. The regions are the squares themselves. The edges of the graph are just the lines between the squares and the vertices of the graph are just the points where the lines on the chessboard intersect. Let's do some counting. Our graph, or chessboard, has e edges. Each region is surrounded by four edges—they *are* called squares after all—and so $r = \frac{1}{2}e$. (You might think this should be $r = \frac{1}{4}e$, but each edge 'belongs' to two regions, one on each side, and so each edge is getting counted twice.) Similarly, each vertex has four edges coming into it, so $v = \frac{1}{2}e$. (Again, each edge has two ends and 'belongs' to two vertices, one at each end.) So, any surface on which we can successfully put this chessboard must have Euler characteristic given by $v - e + r = \frac{1}{2}e - e + \frac{1}{2}e = 0$. But there are only two surfaces with Euler characteristic 0, the torus on the orientable team and the Klein bottle on the non-orientable team!

So, we could try to put a chessboard on a sphere, but it isn't going to fit very well. Something disagreeable will happen to the chessboard. For example, in the very first chapter, we actually did put a chessboard on a sphere in Problem 1.3 when we did a knight's tour on a $2 \times 2 \times 2$ cube. Something strange happened at the corners on this chessboard. A rook at a corner square still has a complete choice among its four usual directions in which to move, but a bishop now has only three options instead of its usual four. That's a little weirder than just the usual thing of having a border for a chessboard. That's not to say we won't ever talk about weird chessboards again—we will—but I hope it does explain why I won't be putting chessboards on projective planes or two-holed tori any time soon.

The Third Dimension

Other than putting chessboards on surfaces such as cylinders and tori as we have been doing, perhaps the most obvious variation for a chessboard would be to make it three dimensional. This certainly occurred to the writers in *Star Trek*, for example. That is to say, you could make a chess 'board' from an $8 \times 8 \times 8$ cube and have the chess pieces occupy individual $1 \times 1 \times 1$ cells

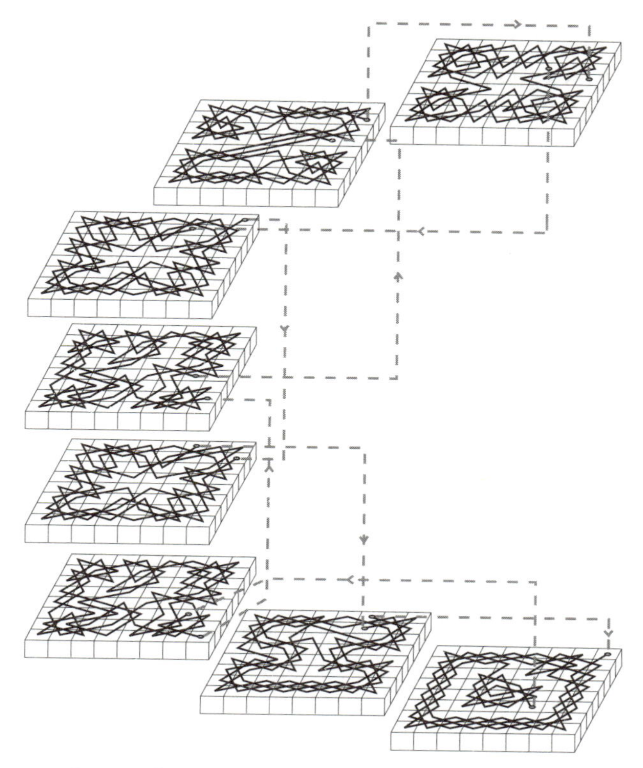

Figure 6.5 A knight's tour of the $8 \times 8 \times 8$ chessboard.

within the cube and then these chess pieces could move within the cube in rather obvious ways from cell to cell. Could a knight tour such a chessboard?

The answer is *yes*, and, in fact, it is not nearly as hard as you might think. Here is the most obvious approach to this three-dimensional knight's tour problem. Think of the $8 \times 8 \times 8$ chessboard as eight regular 8×8 chessboards stacked up, one on top of the other. We know that each individual 8×8 board has a knight's tour. We should be able to splice these together into a single tour in which we tour one layer completely, jump to another layer, tour it completely, jump again, and so on. We just need to arrange the jumps so we get back to the original layer at the very end. This is a bit awkward since a knight can only jump either one or two layers at a time. Here is one order for doing the layers that would work: 1-3-5-7-8-6-4-2-1. This pattern should look familiar, it is exactly the same pattern that we used in Theorem 5.2 to decide in what order to visit the rows of a cylindrical chessboard.

There is still some work to do however: you have to decide which 8×8 tours to use on the individual layers—though this probably doesn't matter very much—and then, more importantly, you have to decide exactly where on each layer to make the links to the connecting layers. This can be a bit tricky, so it is fortunate for us that Ian Stewart has worked out the details of a specific example [29], as shown in Figure 6.5.

Problem 6.2 Find a knight's tour of the $4 \times 4 \times 4$ chessboard.

Here is a delightful three-dimensional chessboard puzzle from Martin Gardner in *New Mathematical Diversions* [15].

Problem 6.3 Imagine a $3 \times 3 \times 3$ cube constructed of 27 individual wooden cubes. Is it possible for a termite to begin on the outside of this cube, to bore into the center of one of the individual $1 \times 1 \times 1$ cubes, and then begin boring its way along always turning only at right angles and always staying parallel to the sides of the cubes—no diagonal boring allowed—and bore through each of the 26 outside cubes once and only once and then finish by boring triumphantly into the center cube? In other words, does the $3 \times 3 \times 3$ chessboard have an open one-step rook's tour ending in the center?

BOXES

I will use the word *box* for chessboards that are on the surface of rectangular solids to distinguish these chessboards from the three-dimensional boards which we have just been discussing. So, using this terminology, in Problem 1.3 we found a knight's tour for a $2 \times 2 \times 2$ box. We can also see that it is easy, though perhaps not very interesting, for a knight to tour a $1 \times 1 \times 1$ box. The question of finding a knight's tour for an $8 \times 8 \times 8$ box was raised well over 200 years ago and presented as a puzzle by H. E. Dudeney in *Amusements in Mathematics* [10]. Dudeney's solution is shown in Figure 6.6, where the idea, to no great surprise,

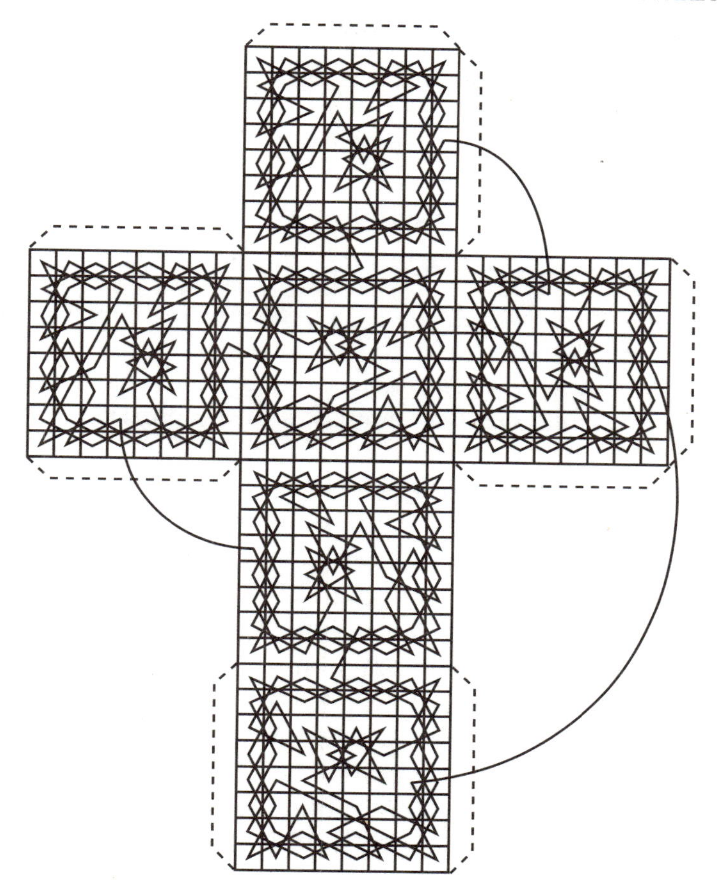

Figure 6.6 Dudeney's tour of the $8 \times 8 \times 8$ box.

is to tour each face separately and then link up these individual tours. A nice additional feature is that Dudeney even added extra little tabs to his drawing so that you could cut out the diagram, fold it into a cube, and tuck in the tabs to hold it all together.

Problem 6.4 Find a knight's tour for the $6 \times 6 \times 6$ box.

Other Variations

Human imagination seems to have no limits whatsoever, and a considerable amount of this imaginative power has been

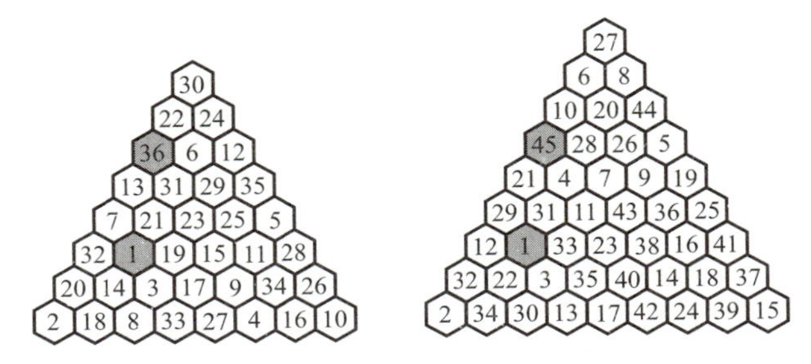

Figure 6.7 Knight's tours of triangular honeycombs.

applied to chessboard problems. There is no particular reason, for example, to restrict ourselves to standard chess moves. In the fourteenth century in Persia they used a chess piece, appropriately called a *camel*, that moves three squares horizontally or vertically and then one square left or right. Euler studied tours of chessboards with camels. But a move consisting of any combination of horizontal and vertical steps could be adopted. This could even be done in three dimensions: perhaps the most 'natural' three-dimensional knight move is three squares in one direction, followed by two squares in another direction, followed by one square in the third direction, rather than the knight's move we used in Figure 6.6.

A particularly fruitful variation of the standard chessboard has recently been invented by Heiko Harborth of the Technical University of Braunschweig in which the geometry of the board itself has changed. He uses hexagons rather than squares and builds chessboards called *triangular honeycombs* in the shape of equilateral triangles. When I first saw these chessboards I immediately thought about doing knight's tours on them and soon found that it was not too difficult to show that any triangular honeycomb chessboard of order 8 or higher has a knight's tour [31]. The idea of the proof was to provide explicit tours for boards having orders 8 through 16. Then, for larger boards, you can subdivide them into four smaller boards in an obvious way and splice together tours of the smaller boards. Tours for boards of orders 8 and 9 are shown in Figure 6.7. In turn, these triangular tours could be used to build tours of hexagonal boards as well.

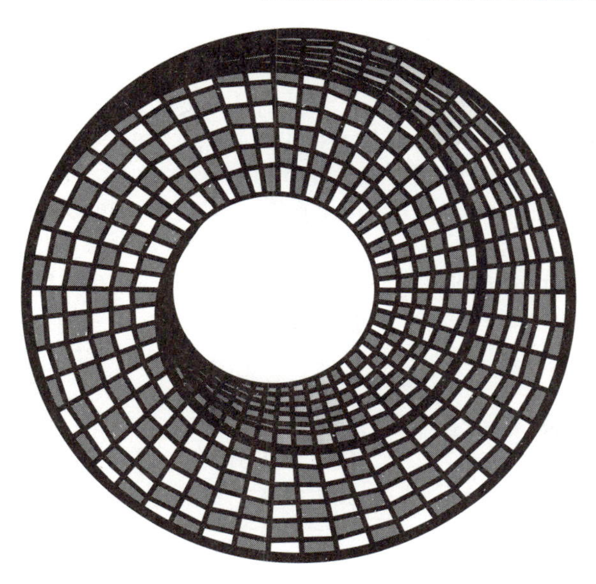

Figure 6.8 A three-dimensional toroidal chessboard with a twist.

The number of variations that would be worth exploring is almost endless, including a general category of ring-shaped boards such as the See's Candies 'chessboard' from Problem 3.2, or general rectangular boxes of dimension $k \times l \times m$, or three-dimensional boards of dimension $k \times l \times m$, or other three-dimensional boards such as a *three-dimensional torus*, which is a solid rectangular block in which all three pairs of opposite faces have been identified, or perhaps even two-dimensional boards placed on surfaces such as the one in Figure 6.8—a square torus with a twist—and three-dimensional boards based on such solid three-dimensional objects, and of course there is no reason to stop with chessboards in only three dimensions!

SOLUTIONS TO PROBLEMS

Solution 6.1 X can win each of these games in a single move by playing as shown in Figure 6.9. Note that for each of these winning diagonal 'three in a row' lines, as they go off of one side, either right or left, and reappear on the other side, not only has the position reversed, but the direction of the diagonal has also changed, which of course makes complete sense when you think about it.

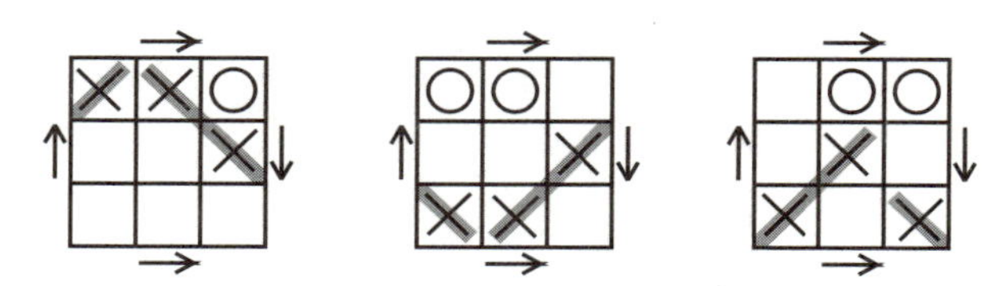

Figure 6.9 X's winning plays.

Solution 6.2 Since there is not even an open tour of the 4×4 chessboard to begin with, this is perhaps a somewhat harder problem than finding a tour for the $8 \times 8 \times 8$ chessboard. Still, using the structure of the knights graph for the 4×4 board exhibited in Figure 3.2, it is possible to cover either four or eight squares each time you visit a layer and to do this in a fairly methodical fashion. One way to do this is shown in Figure 6.10.

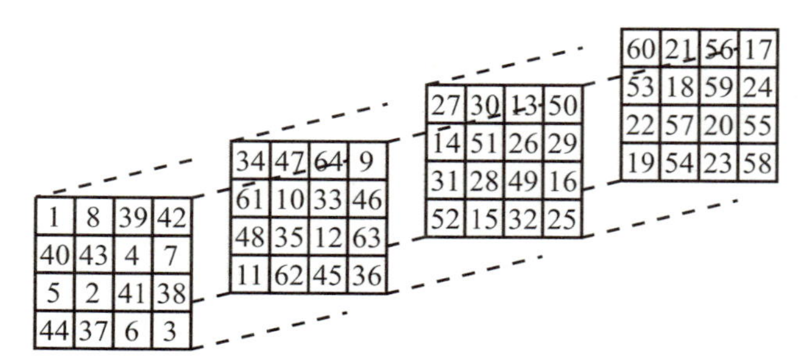

Figure 6.10 A knight's tour of the $4 \times 4 \times 4$ chessboard.

Solution 6.3 The answer is *no*, there is no such tour. Imagine that our $3 \times 3 \times 3$ wooden cube has been carefully crafted from 14 cubes made of maple and 13 cubes made of walnut and put together in an attractive three-dimensional checkerboard pattern. Clearly, then, since the termite is forced to alternate between the two kinds of wood, the termite had better begin on the outside by boring into one of the maple cubes and hope that the center cube is also maple. But the center cube is walnut, and so the termite is doomed to failure, since it can't alternate between 14 maple cubes and 13 walnut cubes and still end on a walnut cube. It is easy to see that this same argument also works, as Gardner points out, for cubes of size $7 \times 7 \times 7$, $11 \times 11 \times 11$, $15 \times 15 \times 15$, and so on.

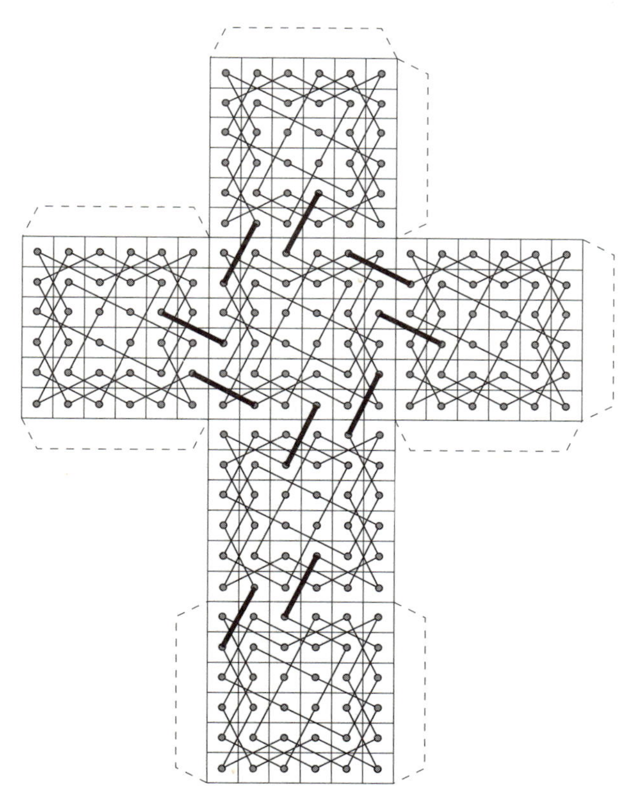

Figure 6.11 A knight's tour of the $6 \times 6 \times 6$ box.

Solution 6.4 One solution found by Arden Rzewnicki and Jesse Howard is shown in Figure 6.11, where extra tabs have been provided if you want to make a three-dimensional model.

Domination

The concept of domination is one of the central ideas in graph theory, and is especially important in the application of graph theory to the real world. Imagine a network of some kind, it could be a communication network such as a cellular phone system or perhaps a network of roads in your local community. Such systems often require vital transmission stations to make them work effectively. A cellular phone company must provide an adequate number of communication links suitably spaced so that customers always have a strong signal for their cell phones. Similarly, your local community needs to provide an adequate number of fire stations suitably spaced so that there is a satisfactory response time to fires anywhere within the community. And, of course, both the phone company and your city council need to do this in a way that is as economical as possible, which means building as few communication links or fire stations as possible.

Dominating Sets

This basic idea of wanting to achieve something quite specific, but as economically as possible, leads quite naturally to the following mathematical definitions.

Definition 7.1 A set S of vertices in a graph G is called a *dominating set* if every vertex in the graph is either in the set S or is adjacent to a vertex in the set S. The *domination number* of a graph G is, then, the minimum size of a dominating set in the graph G. We denote this domination number of the graph G

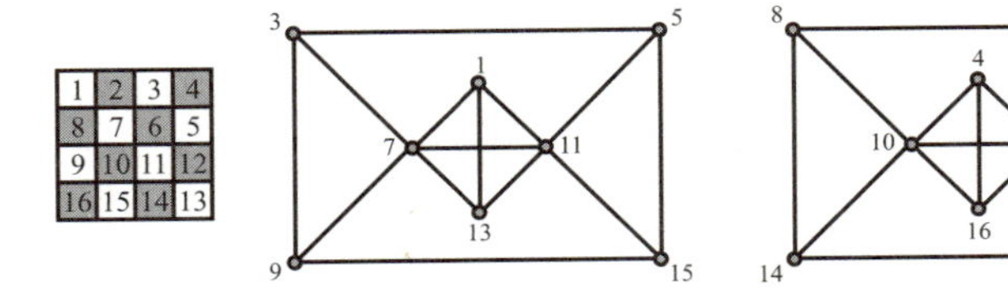

Figure 7.1 The bishops graph $B_{4\times4}$.

by $\gamma(G)$. A dominating set with this minimum size is called a *minimum dominating set*.

We can take as an example the bishops graph $B_{4\times4}$ for the 4×4 chessboard shown in Figure 7.1. Note that this graph is necessarily disconnected because a bishop always stays on squares of a single color. So this graph has two components: one for the white squares, and one for the black squares. The most obvious dominating set for this graph is the set $\{7, 11, 10, 6\}$ and it is quite easy to see at a glance that every vertex in the graph is either in this set or adjacent to one of these four vertices. Moreover, this is a minimum dominating set since clearly neither of the two components of this graph can be dominated by only a single vertex. Therefore, $\gamma(B_{4\times4}) = 4$.

Now, in this case, as it happens, there are other minimum dominating sets for the bishops graph $B_{4\times4}$, for example, $\{6, 7, 14, 15\}$ is also a minimum dominating set. There are also dominating sets, such as $\{3, 5, 6, 10, 13\}$, that share with these *minimum* dominating sets a very important property: if you take any vertex away from them, they will no longer be dominating sets. Such a set is called a *minimal dominating set*. So there is a subtle, but absolutely crucial, distinction in our use of the two words: minim*um* and minim*al*.

The Covering Problem

In the language of chess problems, of course, what we are talking about with the dominating sets in Figure 7.1 is covering the 4×4 chessboard with bishops. The dominating set $\{6, 7, 10, 11\}$ corresponds to placing four bishops in the center of the board so as to cover the entire board. Similarly, covering the board by placing four bishops in a row on squares 9, 10, 11, and 12 corresponds to a dominating set in the bishops graph $B_{4\times4}$. This 'Covering Problem' for bishops on the 4×4 chessboard was quite easy to solve. More generally, we can ask for each chess piece: how many chess pieces of an individual type are required to cover an $n \times n$ chessboard? This is known as *the Covering Problem.* It is among the oldest and most studied problems related to the chessboard, and it and its many variations will occupy our attention for the next several chapters. We begin with our good friend, the knight.

Domination with Knights

Next to the queen, the knight is perhaps the most interesting chess piece upon which to base a study of domination. Even a casual glance at the coverings in Figure 7.2 reveals tantalizing symmetries. It is not hard to confirm, by computer search if all else fails, that these coverings are minimum dominating sets.

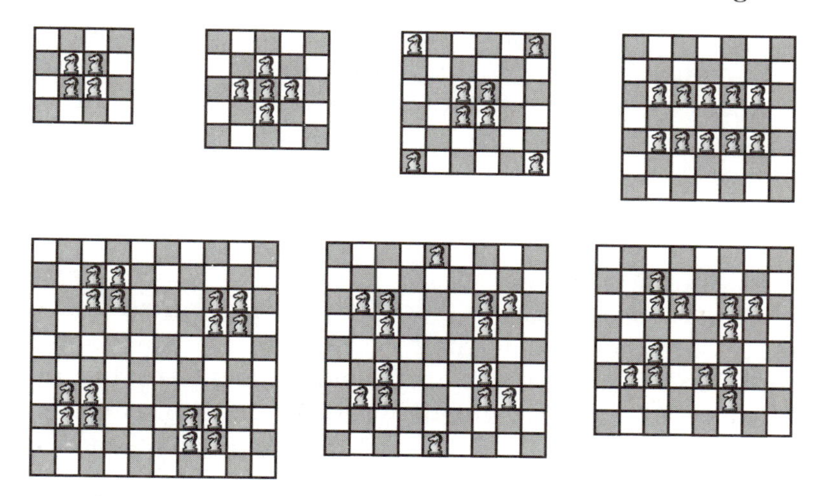

Figure 7.2 Knights domination of chessboards 4×4 through 10×10.

Problem 7.1 Until 1968, the covering in Figure 7.2 for the 10×10 chessboard was believed to be unique. By moving one

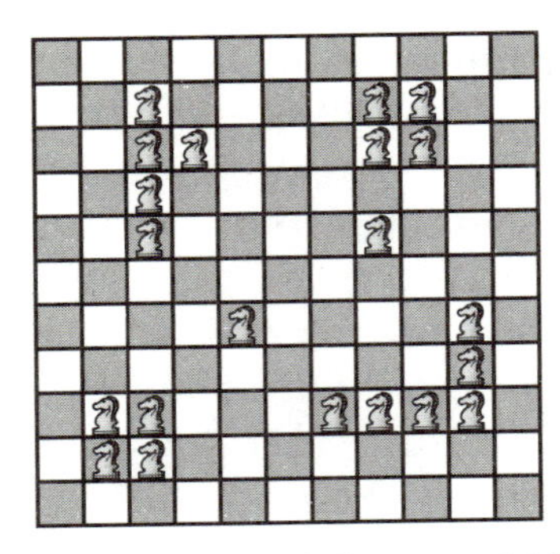

Figure 7.3 A covering of the 11×11 board with 21 knights found in 1971.

knight from each group of four, show that it is easy to find an additional covering—one which is also highly symmetric.

The regularity of the patterns in Figure 7.2 suggests that there might be a hope of solving the Covering Problem for knights in some kind of orderly fashion. Symmetry seems to go completely out of the window, however, for the 11×11 chessboard. In 1971, Bernard Lemaire [21] found the arrangement of 21 knights in Figure 7.3 that covers the 11×11 chessboard and Alice McRae showed that this is best possible. So, unfortunately, analyzing domination for knights remains a very hard problem. On the bright side, however, Eleanor Hare and Stephen Hedetniemi produced a linear-time algorithm in 1987 for finding the domination number of the knights graphs for rectangular chessboards [17].

The coverings in Figure 7.2 do have some rather striking differences. For example, in the coverings for the 9×9 and the 10×10 boards the knights are all independent—that is, none of them attacks another knight; whereas, at the other extreme, in the covering for the 7×7 board the knights all 'guard' one another. This latter situation motivates the following problem of H. E. Dudeney from his *Amusements in Mathematics* [10].

Problem 7.2 There are two possible ways to arrange 7 knights on the white squares of an 8×8 chessboard so as to cover all 32 black squares. Find both of these arrangements. Then, by superimposing and rotating these two arrangements, find the

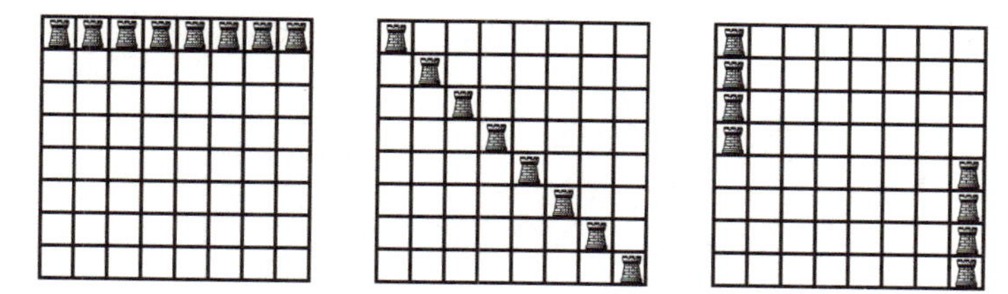

Figure 7.4 Three of the 33,514,312 ways in which 8 rooks can dominate an 8×8 chessboard.

three different ways in which it is possible to place 14 knights on an 8×8 chessboard so that, in Dudeney's words, 'every square shall be either occupied or attacked, and every knight protected by another knight'.

Domination with Rooks

Certainly the chess piece for which domination is the easiest to analyze is the rook, and this was first done by two Russian brothers, Akiva and Isaak Yaglom, who did much of the early work on these types of chessboard problems [40]. For an $n \times n$ chessboard, the rooks domination number is n, that is, $\gamma(R_{n \times n}) = n$, where $R_{n \times n}$ is the rooks graph for the $n \times n$ chessboard. Here is the simple proof. If you have placed fewer than n rooks somewhere on the board in an attempt to cover the board, then there must be a row without a rook in it and, similarly, there must also be a column without a rook in it. Then the square where this empty row and this empty column intersect is clearly not covered by any rook, and so the board itself isn't covered. Hence, you need at least n rooks to cover the board, that is, $\gamma(R_{n \times n}) \geqslant n$. On the other hand, you can place n rooks along a single row, and they obviously dominate the entire board, that is, $\gamma(R_{n \times n}) \leqslant n$. Therefore, $\gamma(R_{n \times n}) = n$.

Problem 7.3 How many different ways can you place n rooks on an $n \times n$ chessboard so that they dominate the board? Here, following Yaglom and Yaglom, two arrangements are to be considered different unless they are identical. For example, there are exactly six ways that two rooks can be arranged to dominate a 2×2 chessboard.

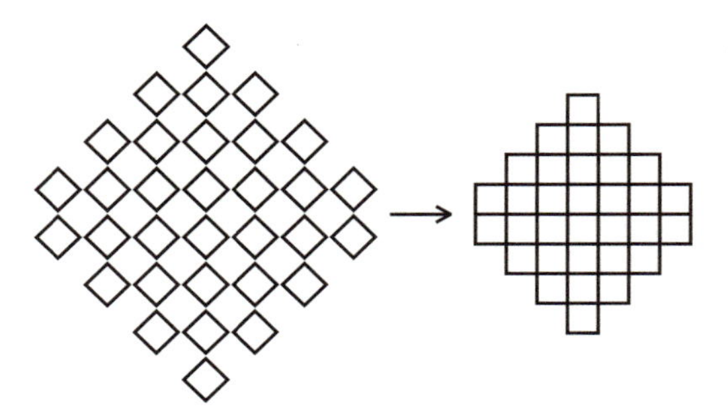

Figure 7.5 White bishops turning into rooks.

Domination with Bishops

Domination with bishops is closely related to rooks domination for the following reason. First of all, as we have already seen, the bishops graph for a chessboard separates naturally into two components: one for the white squares, and one for the black squares. Furthermore, bishops moving on, say, white squares move along white diagonal lines, but if you turn the chessboard—or your head, if you prefer—45° either way, these diagonal lines become either horizontal or vertical lines, and the bishops appear to be moving along these lines just as if they were rooks. In other words, each component of the bishops graph can be thought of as a rooks graph for an appropriate, albeit somewhat irregular, chessboard. It should come as no surprise, then, that for an $n \times n$ chessboard the domination number for bishops is also n, that is, $\gamma(B_{n\times n}) = n$, where $B_{n\times n}$ is the bishops graph for the $n \times n$ chessboard. Here is the proof as given by Yaglom and Yaglom [40].

For n even we will consider the specific case of the 8×8 chessboard, since the general case for n even works in exactly the same way. First, we show that you need at least 8 bishops to cover the board. In Figure 7.5 on the left, we completely remove all the black squares from view and rotate the chessboard clockwise 45°. The bishops now appear to behave like rooks—moving horizontally and vertically—and, since this 'chessboard' contains a 4×5 'board' in the middle, we know that at least 4 bishops will be needed to cover the white squares. If you prefer (and it may make this argument somewhat easier to

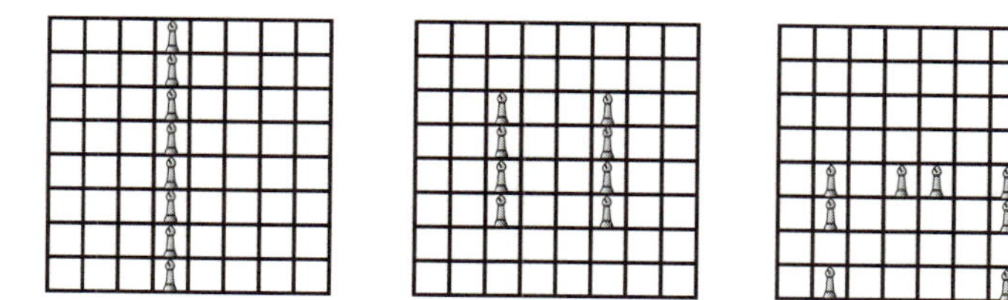

Figure 7.6 Three of the 11,664 ways in which 8 bishops can dominate an 8×8 chessboard.

see), this diagram can be redrawn by putting the white squares into a more rectangular grid-like pattern, as shown on the right in Figure 7.5.

Now, since there is perfect symmetry between the black squares and the white squares on the 8×8 chessboard, at least 4 bishops are also needed for the black squares. Hence, you need at least $4 + 4 = 8$ bishops to cover the 8×8 chessboard, that is, $\gamma(B_{8\times 8}) \geqslant 8$. On the other hand, 8 bishops placed in the fourth column, as shown on the left in Figure 7.6, dominate the entire board, that is, $\gamma(B_{8\times 8}) \leqslant 8$. Therefore, $\gamma(B_{8\times 8}) = 8$, and, in general, for n even, $\gamma(B_{n\times n}) = n$.

For n odd, the argument works the same way, although the convenient symmetry between black and white squares breaks down, so you have to do each color individually. If we write $n = 2k + 1$, then the board corresponding to squares of one of the two colors will contain a $(k + 1) \times (k + 1)$ square and hence need at least $k + 1$ bishops, and the board corresponding to squares of the other color will contain a $k \times k$ square and hence need at least k bishops. You might want to pause to see for yourself how this works for a 7×7 or a 9×9 board. Thus, at least $(k + 1) + k = 2k + 1 = n$ bishops are required for the $n \times n$ board. Finally, n bishops placed in the central column dominate the entire board. So for n odd the domination number is also n. Therefore, for any n, $\gamma(B_{n\times n}) = n$.

Problem 7.4 Show that there are 11,664 different ways you can place 8 bishops on an 8×8 chessboard so that they dominate the board. How many ways are there for 9 bishops on a 9×9 board? Or 10 bishops on a 10×10 board? Here, again, two

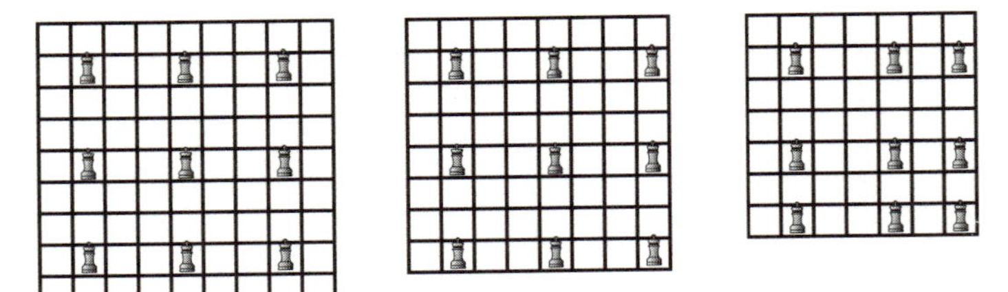

Figure 7.7 Kings dominating the 9×9, 8×8, and 7×7 chessboards.

arrangements are to be considered different unless they are identical.

Domination with Kings

Domination with kings turns out to be surprisingly simple, and even a little boring. The whole story—almost—is told in Figure 7.7, where best-possible coverings of the 9×9, the 8×8, and the 7×7 chessboards are shown. Clearly, since a king covers only the 3×3 grid in which it sits, the arrangement on the left in Figure 7.7 for the 9×9 board is the unique way to cover the 9×9 board with a minimum number of kings. Disappointingly, however, you can't do any better for the 8×8 chessboard than to simply repeat this exact same pattern with 9 kings. It is even more annoying that the 7×7 chessboard still requires 9 kings—you need five kings just to cover the last column and the bottom row, a total of only thirteen squares. How inefficient!

In spite of this apparent inefficiency, however, there is a very nice argument that shows that this is the best that we can hope to do. In Figure 7.8 each of these same three chessboards has been drawn with an identical pattern of 9 squares highlighted on each board. No matter where we might place a king on any of these boards, it will cover only one of these highlighted squares; in other words, a king can't ever cover two of the highlighted squares at the same time. So, on each of these boards, just to get the highlighted squares covered—much less worrying about the whole board—we are going to need at least 9 kings. So, we certainly need at least 9 kings to cover each of these boards.

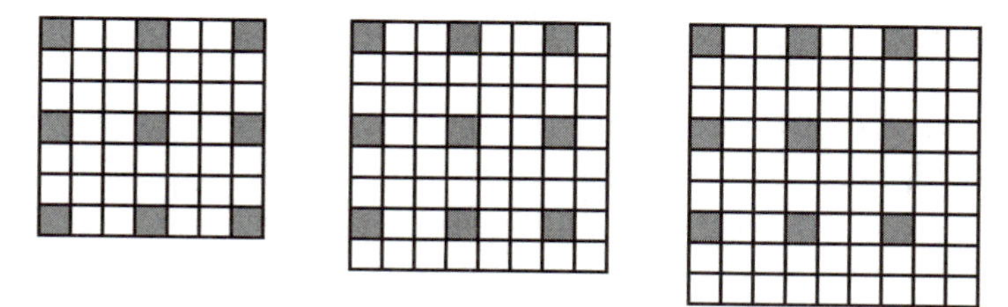

Figure 7.8 Nine highlighted squares that must be covered on each chessboard.

It should be pretty clear now what is going to happen for larger boards. For the 10×10, 11×11, and 12×12 boards, $4^2 = 16$ kings will be required, for the 13×13, 14×14, and 15×15 boards, $5^2 = 25$ kings will be required, and so on. And, in each of these cases, placing these kings in a standard lattice pattern does the job (and, in the case where n is divisible by 3, it is the *only* way to do the job). This can be summarized formally as follows, where $K_{n\times n}$ is the kings graph for the $n \times n$ chessboard, the kings domination number is given by

$$\gamma(K_{n\times n}) = \begin{cases} k^2 = \left(\dfrac{n}{3}\right)^2 & \text{for } n = 3k; \\ (k+1)^2 = \left(\dfrac{n+2}{3}\right)^2 & \text{for } n = 3k+1; \\ (k+1)^2 = \left(\dfrac{n+1}{3}\right)^2 & \text{for } n = 3k+2. \end{cases}$$

These three cases can be combined into a single formula using a convenient function called the floor function, where the *floor* of x, $\lfloor x \rfloor$, is the greatest integer less than or equal to x, in other words, you just 'round down' to the nearest integer. The combined formula for the kings domination number, then, is given by

$$\gamma(K_{n\times n}) = \left\lfloor \frac{n+2}{3} \right\rfloor^2.$$

Note how nicely the floor function works here to give us the same result for $n = 7$, 8, or 9.

Problem 7.5 Find $\gamma(K_{13\times 17})$; that is, find the kings domination number for the 13×17 rectangular chessboard.

SOLUTIONS TO PROBLEMS

Solution 7.1 The covering is shown in Figure 7.9. Note that each of the four knights that were moved made a knight's move to a square that previously only they covered and that, otherwise, each of the four knights that were moved previously covered no other squares that were not already covered by some other knight.

Figure 7.9 Another covering of the 10×10 chessboard.

Solution 7.2 The two arrangements for the 7 knights on white squares to cover all 32 black squares are shown at the top in Figure 7.10. Rotating these patterns 90° clockwise puts the 7 knights on black squares covering all 32 white squares. The two covers for the black squares and the two covers for the white squares can then be superimposed in three combinations, resulting in the three solutions to Dudeney's problem, shown at the bottom in Figure 7.10. The solution on the left is the first arrangement of 7 knights paired with its rotated self. The solution on the right is the second arrangement of 7 knights also paired with itself. And the middle solution is the mixed pairing.

Solution 7.3 In order to count the number of solutions, we note again that in any given solution either all n columns each contain a rook or all n rows each contain a rook—otherwise, some square is uncovered. Since there are n places in which

Figure 7.10 Dudeney's knight guards for the 8×8 chessboard.

to put a rook in a given column, and there are n columns in which to place a rook, the number of ways of arranging n rooks, one in each column, is $n \times n \times \cdots \times n = n^n$. Similarly, the number of ways of arranging n rooks, one in each row, is also n^n. So, the total number of solutions seems to be $n^n + n^n$, but we have to subtract the number of arrangements that we have inadvertently managed to count twice, namely, those for which there is a rook in every column and also a rook in every row, and there are $n!$ of these arrangements since there are n places to put a rook in the first column, and then $n-1$ places to put a rook in the second column, but in a different row than the first rook, and then $n - 2$ places to put a rook in the third column, but in a different row than either of the first two rooks, and so on. Thus, the total number of solutions is $n^n + n^n - n!$, or just $2n^n - n!$. Note that, for $n = 2$, this formula does yield 6 solutions for the 2×2 chessboard. For the ordinary 8×8 chessboard you can check that there are 33,514,312 different solutions.

Solution 7.4

The 8×8 *board.* Since the white and black boards are equivalent we need to do only one color. So let's count the number of ways

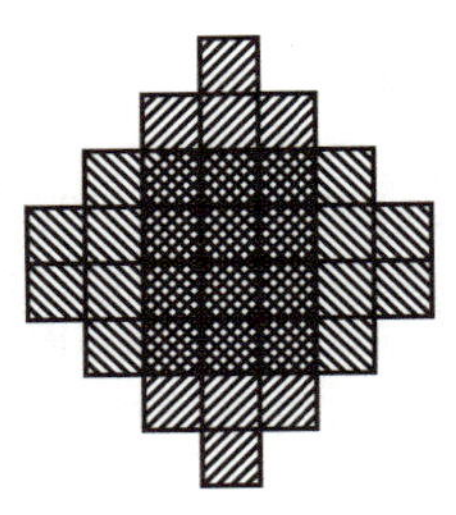

Figure 7.11 The rotated white squares for bishops on the 8×8 board.

to arrange four bishops on the white squares. Again, rotate the board 45° and think of the bishops as rooks. Our objective, then, is to count the number of ways to cover the 'chessboard' shown in Figure 7.11 using just four rooks.

First of all, note that each of the four middle rows of this chessboard must contain a rook since otherwise each of these rows has too many squares to be covered vertically by just four rooks. Similarly, and for the same reason, there must be a rook in each of the three middle columns. Conversely, it is clear that if each of the four middle rows and each of the three middle columns contains a rook, then the board will be completely covered.

There are two cases to consider: either all four rooks are in the three middle columns, or else one rook of the four is not in the three middle columns. In the first case, we are going to place four rooks in three columns in such a way that there is at least one rook in each column. Of course, then, one column must get two rooks. So, first of all, we have 3 choices of which column gets the two rooks. Then, within that column, there are 4 squares in which to put the 2 rooks, and this can be done in 6 different ways—that is, in '4 choose 2' or $\binom{4}{2} = \frac{4\times3}{2} = 6$ ways. Now there remain only two rows and two columns for the last two rooks, and these rooks can be placed in only $2! = 2$ ways. This makes a total of $3 \times 6 \times 2 = 36$ different arrangements that are possible in this first case.

In the second case, there are 12 squares outside of the three middle columns where we can place a rook. Now, having placed that rook in one of those 12 squares, there are three rows still available, and so there are $3! = 6$ ways to place the three remaining rooks in three columns and three rows. This makes a total of $12 \times 6 = 72$ different arrangements that are possible in this second case.

So, there are $36 + 72 = 108$ ways to place four white bishops on an 8×8 chessboard to cover all the white squares and also, therefore, 108 ways to place four black bishops on an 8×8 chessboard to cover all the black squares. Since any arrangement of the white bishops can be combined with any arrangement of the black bishops to yield an arrangement of eight bishops that covers the entire chessboard, there are $108^2 = 11{,}664$ ways to cover the 8×8 board with eight bishops.

The 10×10 *board.* This board can be done in the same manner as the 8×8 board. The rotated board for the white squares is shown in Figure 7.12. Here, each of the five middle columns and each of the four middle rows must contain at least one of the five rooks. Again, there are two cases: either all five rooks are in the four middle rows, or else one rook of the five is not in the four middle rows. In the first case, there are 4 rows from which to choose a row in which to place the two rooks, and within that chosen row there are 5 squares in which to put the 2 rooks, and this can be done in 10 different ways—that is, in '5 choose 2' or $\binom{5}{2} = \frac{5 \times 4}{2} = 10$ ways. The three remaining rooks can now be placed in the three remaining columns and the three remaining rows in $3! = 6$ ways. This makes a total of $4 \times 10 \times 6 = 240$ different arrangements that are possible in this first case.

In the second case, there are 18 squares outside the four middle rows where we can choose to place a rook, which uses up one of the columns. Then, there are $4! = 24$ ways to place the four remaining rooks in the four middle rows and the four remaining middle columns. This gives us $18 \times 24 = 432$ arrangements that are possible in this second case. So there are a total of $240 + 432 = 672$ ways to place five white bishops and, hence, $672^2 = 451{,}584$ total arrangements for covering the 10×10 board with ten bishops.

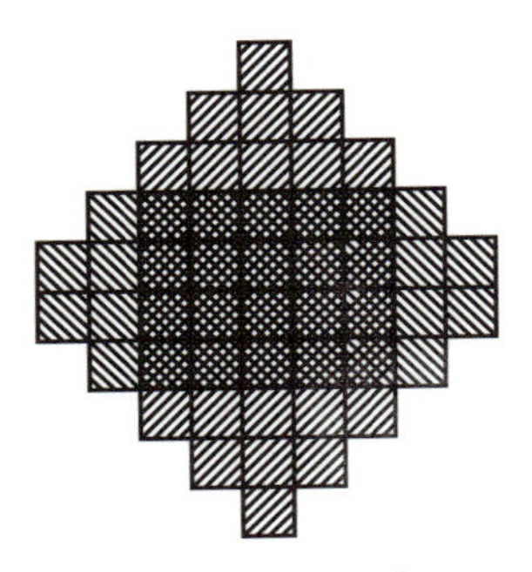

Figure 7.12 The rotated white squares for bishops on the 10×10 board.

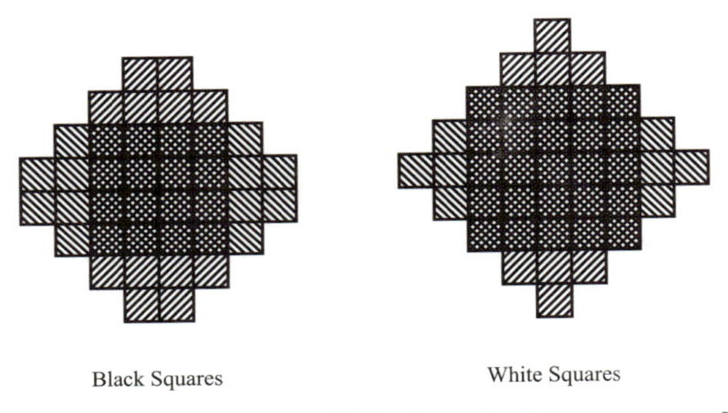

Figure 7.13 The rotated black and white squares for bishops on the 9×9 board.

The 9×9 *board.* We may as well assume that the four corner squares are white. The rotated black board and the rotated white board are then as shown in Figure 7.13. We need to count the numbers of ways to place four rooks on the black board, and five rooks on the white board. But on the black board, each of the four middle rows must contain a rook and each of the four middle columns must also contain a rook, and so we can place the four rooks on this board in exactly $4! = 24$ ways.

The white board is trickier. For the white board, the most we can say is that either the five middle rows are each occupied by a rook, or the five middle columns are each occupied by a rook, or possibly both. Let us count the number of arrangements in which each of the five middle rows are occupied by a rook. In this situation we need to take care that the middle three columns are also each occupied by a rook. There are several cases to consider.

The first case has all five rooks in the middle three columns. This can be done with three rooks in a single column in $3\times10\times2 = 60$ ways: that is, 3 choices for the column, 10 choices for where to put 3 rooks in 5 squares, that is, '5 choose 3' or $\binom{5}{3} = \frac{5\times4}{2} = 10$, and then 2 choices for placing the last two rooks in the remaining two rows and two columns. Or, this can be done with two rooks in each of two of the three columns in $3 \times 10 \times 3 = 90$ ways: that is, 3 choices for which column gets only a single rook, 10 choices within the first of the other two columns for where to put the first 2 rooks in 5 squares, that is, '5 choose 2' or $\binom{5}{2} = \frac{5\times4}{2} = 10$, and then 3 choices for where to put the remaining 2 rooks in the last column in 3 squares, that is, '3 choose 2' or $\binom{3}{2} = 3$. Thus, the first case has $60+90 = 150$ different arrangements.

The second case has four rooks in the middle three columns, two of which must be in the same column. This can be done in $18 \times 3 \times 6 \times 2 = 648$ ways: that is, 18 choices for a square outside the middle three columns in which to place the fifth rook (and this uses up a row), 3 choices of which middle column gets two rooks, 6 choices for where to put 2 rooks in 4 squares (only four rows are available), that is, '4 choose 2' or $\binom{4}{2} = \frac{4\times3}{2} = 6$, and then 2 choices for placing the last two rooks in the remaining two rows and two columns. Thus, the second case has 648 different arrangements.

Finally, the third case has three rooks in the middle three columns. Since there are 18 squares outside the middle three columns, the number of ways to place 2 rooks outside of the middle three columns is '18 choose 2' or $\binom{18}{2} = \frac{18\times17}{2} = 153$, but from this we must subtract the number of arrangements that have two rooks in the same row, which, row by row for the five middle rows, is 1, 6, 15, 6, and 1, or 29 in all, that is,

$$1 = \text{'2 choose 2'} = \binom{2}{2},$$

$$6 = \text{'4 choose 2'} = \binom{4}{2} = \frac{4\times3}{2},$$

and

$$15 = \text{‘6 choose 2’} = \binom{6}{2} = \frac{6 \times 5}{2}.$$

This gives us a total of $153 - 29 = 124$ ways to place two rooks outside of the middle three columns, but in different rows. The remaining three rooks can then be placed in the middle three columns and the remaining three rows in $3! = 6$ ways. Thus, the third case has $124 \times 6 = 744$ different arrangements.

Putting this all together, the total number of ways in which the middle five rows are each occupied is $150 + 648 + 744 = 1542$. Similarly, the total number of ways in which the middle five columns are each occupied is also 1542. The total number of arrangements, then, for the white board is $1542 + 1542 - 120 = 2964$, since we must subtract $5! = 120$ for any arrangements that have been counted twice, that is, any arrangements in which both the middle five rows and the middle five columns have five rooks.

At long last, we can recombine the black board and the white board, and we get $24 \times 2964 = 71{,}136$ total arrangements for covering the 9×9 board with nine bishops.

I'd like to add that the Yaglom brothers carried out these counting arguments in general and have provided the following formulas for the number of different ways you can place n bishops on an $n \times n$ chessboard so that they dominate the board. There are four cases.

Case 1: $n = 4k$, # ways $= \left(\frac{(2k)!(4k+1)}{2}\right)^2$.

Case 2: $n = 4k+1$, # ways $= (2k)!(2k)!\frac{16k^3 + 24k^2 + 11k + 1}{2}$.

Case 3: $n = 4k + 2$, # ways $= ((4k^2 + 5k + 2)(2k)!)^2$.

Case 4: $n = 4k + 3$, # ways $= (2k + 1)!(2k)!(16x^4 + 56x^3 + 67x^2 + 33k + 6)$.

Note that, just as we would expect, in the two even cases, the number of arrangements is a perfect square.

Solution 7.5 $\gamma(K_{13\times17}) = 30$. The arrangement of 30 kings shown on the top in Figure 7.14 shows that $\gamma(K_{13\times17}) \leqslant 30$. The pattern of 30 highlighted squares shown on the bottom in Figure 7.14, each of which requires a different king in order to be covered, shows that $\gamma(K_{13\times17}) \geqslant 30$.

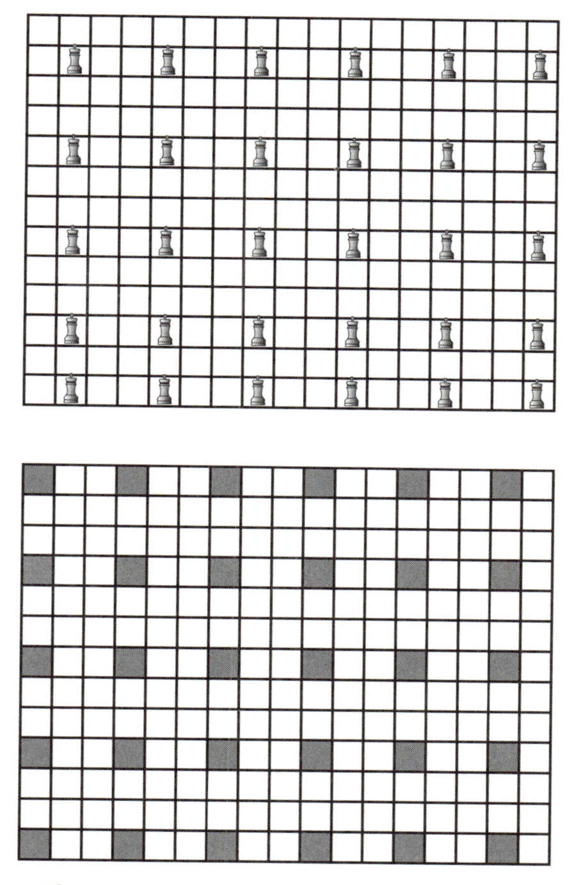

Figure 7.14 Proof that $\gamma(K_{13\times17}) = 30$.

Queens Domination

Of all the chessboard-domination problems, it is that of the queen that continues to hold the most interest among mathematicians. This seems to be a remarkably difficult problem and one that is far from solved even today, although there is much that is known. For example, as we noted in Chapter 1, five queens are required to dominate the 8×8 chessboard, and it has been observed by Yaglom and Yaglom that there are exactly 4860 different ways that these five queens can be arranged so as to dominate the board. Several of these arrangements are shown in Figure 8.1.

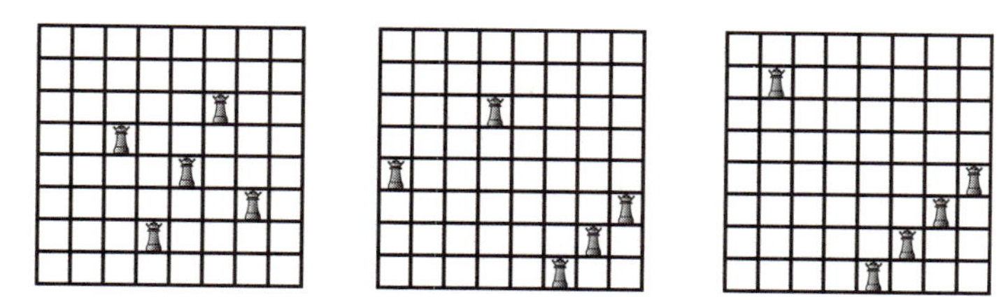

Figure 8.1 The 8×8 chessboard dominated by 5 queens.

Diagonal Domination

Problem 8.1 One of the most interesting arrangements of five queens dominating the 8×8 chessboard is one that has all five placed along the main diagonal. Find such an arrangement. Similarly, find an arrangement of five queens that dominate the 8×8 chessboard with all five placed along a sub-diagonal, that is, along a diagonal line one step above or below the main diagonal.

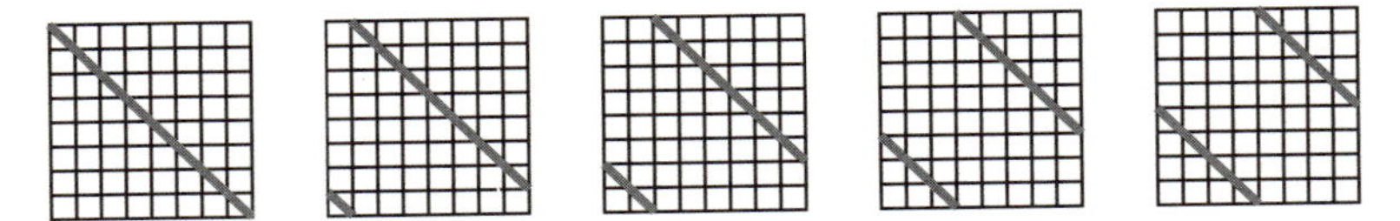

Figure 8.2 The five different *toroidal* diagonals.

Then, extend this idea to 'sub' sub-diagonals by using what I'll call *toroidal* diagonals, that is, 'diagonals' formed as if the chessboard were on a torus. A chessboard has essentially five of these toroidal diagonals, as shown in Figure 8.2. You have already placed queens on the first two of these toroidal diagonals. For which of the remaining three toroidal diagonals in Figure 8.2 is there an arrangement of five queens along the 'diagonal' that dominates the 8×8 chessboard? Note that two of the arrangements of five queens in Figure 8.1 are almost, but not quite, such toroidal diagonal solutions.

Finally, is it possible to place five queens in a single column to dominate the 8×8 chessboard?

The rather surprising 'diagonal' solution to the domination problem for the 8×8 chessboard mentioned in Problem 8.1 is so unexpectedly nice that it naturally makes one wonder whether the queens domination number for an $n \times n$ chessboard could always be achieved by placing the required number of queens along the main diagonal. Also, it is quite natural for us to define the *queens diagonal domination number* to be

$$\operatorname{diag}(Q_{n\times n}) = \left\{\begin{array}{l}\text{the minimum number of queens all}\\ \text{placed along the main diagonal required}\\ \text{to cover the } n \times n \text{ chessboard}\end{array}\right\},$$

and then it is natural for us to study this number in its own right.

Ernie Cockayne and Stephen Hedetniemi have done just that and, in fact, have been able to characterize $\operatorname{diag}(Q_{n\times n})$ in terms of a well-studied number-theoretic function involving arithmetic progressions [8]. One of their conclusions is that once n is sufficiently large, you will definitely need more queens in order to dominate a chessboard if you insist on placing them along the main diagonal than you would if you are willing to place

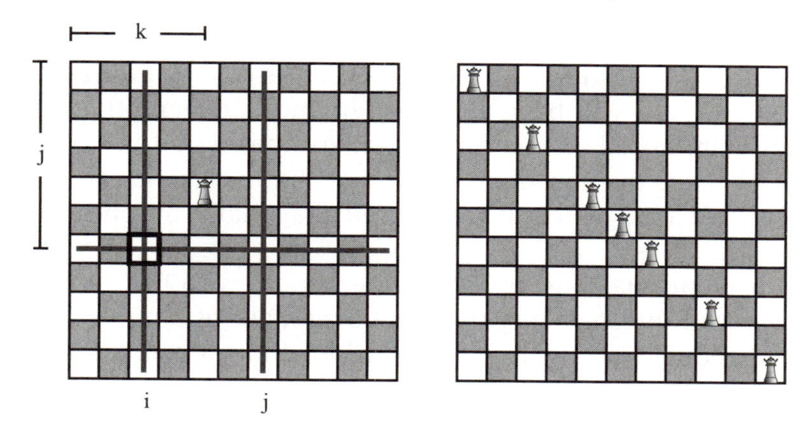

Figure 8.3 Diagonal dominating sets.

them anywhere on the board. So, as we might have guessed, what was true for the 8×8 board was simply too good to last for larger chessboards.

Let's take a look at their basic idea. Suppose you have an arrangement of queens placed along the main diagonal of an $n \times n$ chessboard that dominates the board. Note, first, that in Figure 8.3 we follow the usual tradition in chess of having the lower right-hand square be white, and so the main diagonal consists of all white squares. Now let C be the set of columns that do not contain any queens. For any two columns i and j in C, the corresponding square in the ith column and in the jth row down must be white since this square can only be covered by a queen diagonally. Therefore, $i + j$ is even. This means that either all the numbers in C are even, or else they are all odd. Furthermore, the queen that is covering this square diagonally is in some column k, and in the kth row down, and so $i + j = k + k$; that is, $k = \frac{1}{2}(i + j)$. This means that whenever columns i and j are empty, the column that is in the exact middle between i and j is not empty. Therefore, the set C has the following additional property: for any two numbers in C, their 'midpoint' is not in C; such a set of numbers is called *mid-point free*.

Now, it is easy to see that this idea completely reverses. In other words, if C is a mid-point free subset of $\{1, 2, 3, \ldots, n\}$ consisting only of even numbers, or else, only of odd numbers, then by placing queens along the main diagonal of an $n \times n$ chessboard in all columns except those columns in the subset C, you will have an arrangement of queens that dominates the board. If you have a square that is not covered by one of these queens along a row or column, that square must be white, and, hence, it is covered diagonally by one of the queens.

Since our goal is to cover the chessboard with the minimum number of queens along the main diagonal, we can now recast the diagonal queens domination problem as a problem about numbers:

$$\operatorname{diag}(Q_{n\times n}) = n - \left\{\begin{matrix}\text{maximum size of a mid-point free, all} \\ \text{even or all odd, subset of } \{1,2,3,\ldots,n\}\end{matrix}\right\}.$$

For example, for $n = 8$, we can let $C = \{2,6,8\}$—check that this set is mid-point free—and place 5 queens along the main diagonal in columns 1, 3, 4, 5, and 7. This is the arrangement shown on the left in Figure 8.15. For $n = 11$, we can let $C = \{2,4,8,10\}$ and place queens in columns 1, 3, 5, 6, 7, 9, and 11. This arrangement is shown in Figure 8.3. Thus, $\operatorname{diag}(Q_{11\times 11}) = 7$. But, we will see later that the 11×11 chessboard can be covered by just 5 queens, and so, even as early as $n = 11$, the domination number and the diagonal domination number differ.

The Spencer–Cockayne Construction

Now, the very first arrangement of five queens given in Figure 8.1 also happens to be especially appealing, largely due to its symmetry. It seems somewhat unlikely that such a nice pattern could be extended to larger boards, but in fact there is a beautiful construction due to Spencer and Cockayne that tries to do just that [6]. Here is how it goes.

Begin with a single queen on a square somewhere in the middle of a very large chessboard. You can think of the board as an infinite chessboard if you like or else just think of it as a board that is simply so large we aren't likely to ever reach one of its edges anytime soon. Now this single queen dominates lots of

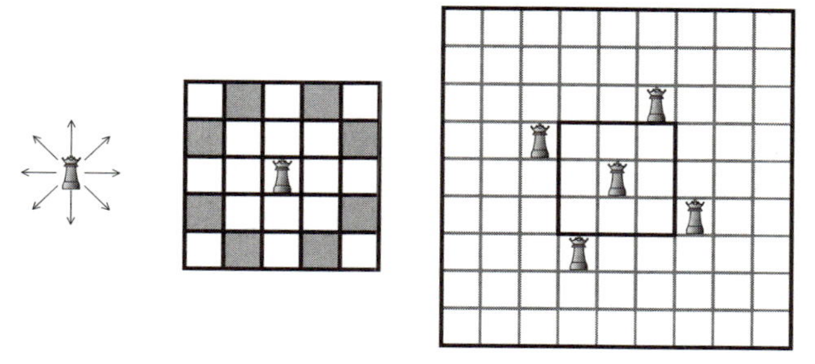

Figure 8.4 The first stage of the Spencer–Cockayne construction.

squares in all directions, but in particular, it dominates the nine squares of the 3×3 square in which it sits. However, if we look at the 5×5 square in which this queens also sits, as shown in Figure 8.4, we see that there are now eight shaded squares that are not covered by this queen. You also can't help but notice—and I believe this is no accident—that these shaded squares are all a knight's move from the queen! In other words, the knight's move in chess has evolved over the centuries to be, in this very specific sense, dual to the queen's move.

Now look what happens when we place four additional queens in four of these shaded squares in the symmetric fashion shown on the right in Figure 8.4. These 5 queens now control a 9×9 chessboard! And, hence, of course, they also control the 8×8 chessboard contained within it, just as we saw in Figure 8.1. The second stage of the construction is shown in Figure 8.5. In the 11×11 chessboard that now surrounds the 9×9 chessboard, there are again eight shaded squares that are not covered by the 5 queens currently on the board. But, if we again place four more queens in four of these shaded squares in the same symmetric fashion as before, these 9 queens now control a 15×15 chessboard!

Problem 8.2 We are now off and running with the Spencer–Cockayne construction, adding four queens at a time to control ever larger square chessboards. But the construction does contain a hidden surprise. Do several more iterations of the construction until you discover what the surprise is.

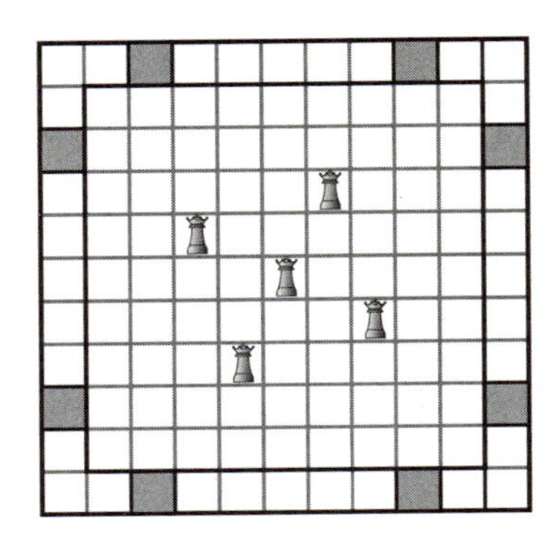

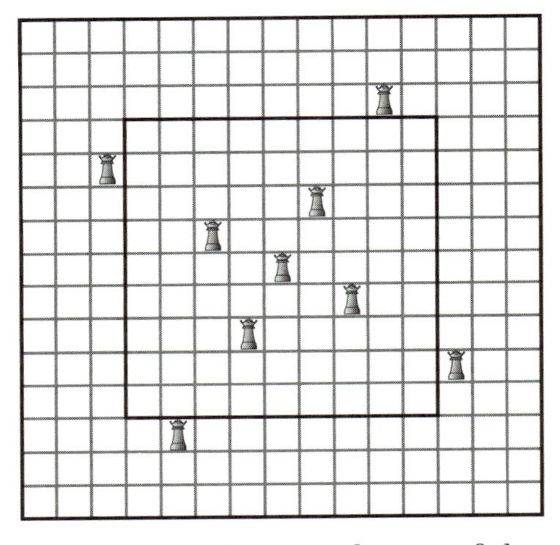

Figure 8.5 The second stage of the Spencer–Cockayne construction.

At this point you may be beginning to think that the queens domination problem isn't so hard after all. But wait: we don't yet know whether the arrangements of queens produced by the Spencer–Cockayne construction for various chessboards are optimal or not. They seem to be so efficient that it is certainly hard to imagine being able to do any better. In this construction the queens are really just perfectly aligned to catch all the diagonals needed to cover the corner areas of the board and yet, at the same time, catch all the rows and columns needed to cover the middle section of the board at both the top and bottom and on the left and right. You can think of this construction as working like a big wide paintbrush used to paint a square with four strokes: one horizontal stroke, one vertical stroke, and then one stroke in each of the two diagonal directions. This is why the queens can cover a square of a given size but then the next size up has eight little squares left symmetrically uncovered—they are the tiny gaps that are just barely missed by the paint brush.

As it happens, we don't actually know whether the arrangement of 9 queens covering the 15×15 board in Figure 8.5 is in fact an optimal covering or not, but I might as well tell you now that, sadly, at the very next stage of the construction, the 21×21 board can definitely be covered by fewer than the 13 queens that are produced by the Spencer–Cockayne construction. Remember, I did say that queens domination was a difficult problem. This is very strong evidence of that fact. So, let's now turn our attention to thinking about ways to decide whether a given arrangement of queens might be optimal or not. Along the way we will want to do two things. We will want to build a short table of values that gives the queens domination numbers for $n \times n$ chessboards for the first dozen or so values of n, and we will also want to describe both upper and lower bounds that have been found that can at least give us some information about the size of the queens domination number for a given value of n even if we don't know this value exactly. Throughout, we will follow a treatment given by Weakley in his paper 'Domination in the queen's graph' [36].

AN UPPER BOUND

Let's begin by finding an upper bound for the number of queens needed to cover an $n \times n$ board.

Theorem 8.1 (Welch) *For* $n = 3m + r, 0 \leqslant r < 3$:

$$\gamma(Q_{n\times n}) \leqslant 2m + r.$$

Welch's easy—well, at least it's easy once someone clever thinks of it—construction is to divide a $3m \times 3m$ chessboard

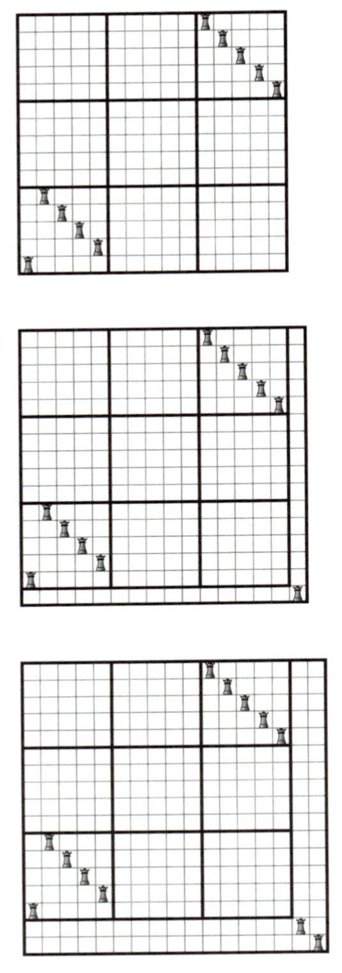

Figure 8.6 Covering a board with two-thirds of the queens, plus an extra queen or two.

into nine $m \times m$ squares. He then puts m queens in the upper right-hand $m \times m$ square, and m queens in the lower left-hand square in such a way that the entire $3m \times 3m$ board is covered. Note the clever way all the squares in the center are covered along diagonals. Any leftover r rows and columns can then be covered by just r queens placed in the corner. This idea is illustrated in Figure 8.6.

Note, for example, that for the 15×15 board the upper bound given by Welch's theorem is 10, which is actually fairly close to 9, the number of queens with which we have been able to cover the 15×15 board in Figure 8.5.

Since it is obviously possible to cover an $n \times n$ chessboard with n queens, we can think of Theorem 8.1 as saying that roughly two-thirds of that number of queens is enough to do the job. On the other hand, the next theorem will tell us, again roughly, that we do need at least half that many queens; that is, this next theorem will provide us with a lower bound for the queens domination number. So, we will soon have the queens domination number trapped somewhere between roughly a half and two-thirds of n. The rather remarkable lower bound is due to Spencer, and the proof presented here will follow a version given by Weakley. The following easy problem will play a key role in the proof.

Problem 8.3 Show by an explicit construction that $\gamma(Q_{n\times n}) \leqslant n - 2$. That is, find a way to place $n - 2$ queens on an $n \times n$ chessboard so as to cover the board.

Spencer's Remarkable Lower Bound

Theorem 8.2 (Spencer) $\gamma(Q_{n\times n}) \geqslant \frac{1}{2}(n-1)$.

Proof. We begin by assuming that we already have a minimum number of queens placed on an $n \times n$ chessboard that dominate the board; that is, we have $\gamma = \gamma(Q_{n\times n})$ queens placed on the chessboard, and they cover the board. Because of Problem 8.3, we know that $\gamma \leqslant n-2$. A key observation that follows immediately from this, then, is that there must be at least two rows that have no queens in them at all and, also, at least two columns that have no queens.

So, having said that, it makes sense for us to now let a be the leftmost empty column, and to let b be the rightmost empty column, and also for us to let c be the lowest empty row, and to let d be the highest empty row, as shown in Figure 8.7. Also, we may as well assume that $b - a \geqslant d - c$. If not, we can just rotate the chessboard 90° and appropriately relabel these same four empty rows and columns.

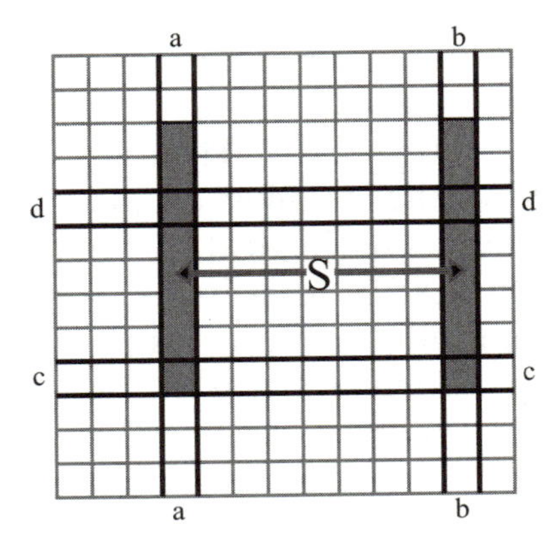

Figure 8.7 The proof of Spencer's Theorem.

Next, we let S be the set of shaded squares shown in Figure 8.7 consisting of the $b - a$ consecutive squares in column a starting at the bottom with square (a, c), together with the $b - a$ consecutive squares in column b starting at the bottom with square (b, c). Because $b - a \geqslant d - c$, it is guaranteed that the set S contains all the squares in columns a and b that are strictly between rows c and d. In fact, it was not important exactly where the shaded squares in S were placed as long as they ended up including those squares in columns a and b strictly between rows c and d. And, of course, our set S contains exactly $2(b - a)$ squares.

Note that the height of the set S was very carefully chosen so that any diagonal on the chessboard will intersect the set S in at most one square. Therefore, a queen in any row above or below S controls at most two squares of S and, since each such row is non-empty—remember that c is the lowest empty row and d is the highest empty row—there must be at least

$$(c - 1) + (n - (c + (b - a))) = n - (b - a) - 1$$

such queens. Note: we had to be cautious here because in the special case that $b - a = d - c$, the first row immediately above the shaded squares of S would in fact be row d, which we know has no queens, so we were careful to count only the $(n - (c + (b - a)))$ rows above S that we could be completely sure contained queens rather than the $(n - (c + (b - a) - 1))$ actual rows above S.

Any remaining queens—and there are a total of γ queens on the board—can control at most four squares of S. Therefore, since all $2(b - a)$ squares in S must be controlled by some queen,

$$2(n - (b - a) - 1) + 4(\gamma - (n - (b - a) - 1)) \geqslant 2(b - a),$$

and, solving for γ, this rather amazingly reduces to $\gamma \geqslant \frac{1}{2}(n - 1)$, which is Spencer's lower bound, thus completing the proof of the theorem. □

I've Got a Little List

It is high time we begin to produce some actual queens domination numbers for specific values of n. It is of course obvious that $\gamma(Q_{n\times n}) = 1$ for $n = 1$, 2, and 3. We can also see from Figure 8.4 that 1 queen is not sufficient to cover a 4×4 chessboard, but that 2 queens will be. Alternatively, however, we can apply Theorem 8.2, when $n = 4$, which gives $\gamma(Q_{4\times4}) \geqslant \frac{1}{2}(4 - 1) = \frac{3}{2}$. But since γ is an integer this means that, in reality, $\gamma(Q_{4\times4}) \geqslant 2$. So, then, the covering for the 4×4 chessboard shown in Figure 8.8 demonstrates that $\gamma(Q_{4\times4}) = 2$.

For $n = 5$, Figure 8.4 shows that if a first queen is placed in the center, then there is no placement for a second queen that will cover the entire board. It is easy to check that this is also true for any of the other five essentially different squares on which the first queen could be placed. Thus, at least 3 queens are required. One covering for the 5×5 chessboard with 3 queens is shown in Figure 8.8, and so $\gamma(Q_{5\times5}) = 3$. Note that Theorem 8.2 isn't particularly helpful here since the lower bound it produces is $\frac{1}{2}(5 - 1) = 2$.

For $n = 6$, Theorem 8.2 gives $\gamma(Q_{6\times6}) \geqslant \frac{1}{2}(6 - 1) = \frac{5}{2}$, which again means that, in reality, $\gamma(Q_{6\times6}) \geqslant 3$. So, with this as a

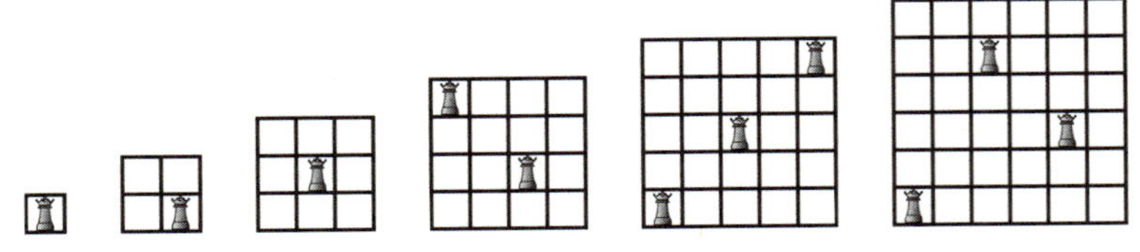

Figure 8.8 The first six queens domination numbers.

lower bound, the covering for the 6×6 chessboard shown in Figure 8.8 tells us that $\gamma(Q_{6\times6}) = 3$.

Now, as we have seen several times, in the case where n is an even number, Spencer's lower bound can technically never be achieved because $\frac{1}{2}(n-1)$ isn't even an integer, so in fact the lower bound given by Spencer's theorem in reality becomes $\frac{1}{2}n$ whenever n is even. So, for example, this revised lower bound was achieved for $n = 2, 4$, and 6, and, as we shall see, it will be achieved for quite a few other values of n when n is even.

What about the case when n is odd? Can the lower bound be achieved? Well, the lower bound was achieved in the case $n = 3$, since $\gamma(Q_{3\times3}) = 1 = \frac{1}{2}(3-1)$. But, then, it wasn't achieved in the case $n = 5$, as we just saw. The following theorem tells us that this was no accident, it turns out that the only values of n for which we have a chance of achieving Spencer's lower bound are those that give us a remainder of 3 upon division by 4, that is, for $n = 3, 7, 11$, and so on. Therefore, for $n = 5, 9, 13$, and so on, the lower bound is about to be raised! But it is going to take some work. A lot of work.

Weakley's New Improved Lower Bound

Theorem 8.3 (Weakley) *If the lower bound of Theorem 8.2 is attained for a given $n \times n$ chessboard, that is, if $\gamma(Q_{n\times n}) = \frac{1}{2}(n-1)$, then $n \equiv 3 \bmod 4$.*

Proof. We begin as in the proof of Theorem 8.2 by assuming that we already have a minimum number of queens placed on an $n \times n$ chessboard that dominate the board; that is, we have $\gamma = \gamma(Q_{n\times n}) = \frac{1}{2}(n-1)$ queens placed on the board, and they cover the board. We continue, again as was shown in Figure 8.7, by defining rows a and b, columns c and d, and the set S to be exactly as in the proof of Theorem 8.2, and by once again assuming, as we did there, that $b - a \geqslant d - c$.

Let i be the number of queens that are either in a row above S or in a row below S. Since these queens can cover at most two squares of S, and any of the other $\gamma - i$ queens can cover at most four squares of S, and since S has $2(b - a)$ squares in all, we have that

$$2(b - a) \leqslant 2i + 4(\gamma - i).$$

But, by assumption, $\gamma = \frac{1}{2}(n - 1)$, and so this inequality reduces easily to

$$i \leqslant n - (b - a) - 1.$$

Also, by definition, all rows above S and all rows below S are non-empty and contain queens, except possibly row d, so

$$i \geqslant n - (b - a) - 1.$$

Thus, $i = n - (b - a) - 1$. So, we can conclude that above d and below c there is exactly one queen per row. We can also conclude that $d - c = b - a$, since, if $d - c < b - a$, then $i \geqslant n - (b - a)$, a contradiction.

Now, let $k = (b - a) + 1$, and so $i = n - k$, and we can draw a square $k \times k$ sub-board B whose edges are formed by columns a and b, and rows c and d, as shown in Figure 8.9, and each row above or below the square sub-board B contains one and only one queen. Similarly, each column to the left or right of B contains one and only one queen.

Now, if we back up a couple of paragraphs to the inequality $2(b - a) \leqslant 2i + 4(\gamma - i)$ and replace γ by $\frac{1}{2}(n - 1)$ and also replace i by $n - (b - a) - 1$, we see that what started out there as an *in*equality has now become an equality, since the right-hand side of this expression reduces to $2(b - a)$. This means that, in fact, each of the i queens above or below B controls exactly two squares of S. In particular, then, each of these queens lies strictly between columns a and b and controls one square of S along a positive diagonal, and one square of S along a negative diagonal.

Let E be the set of squares in columns a and b and rows c and d forming the outer edge of the square sub-board B, as

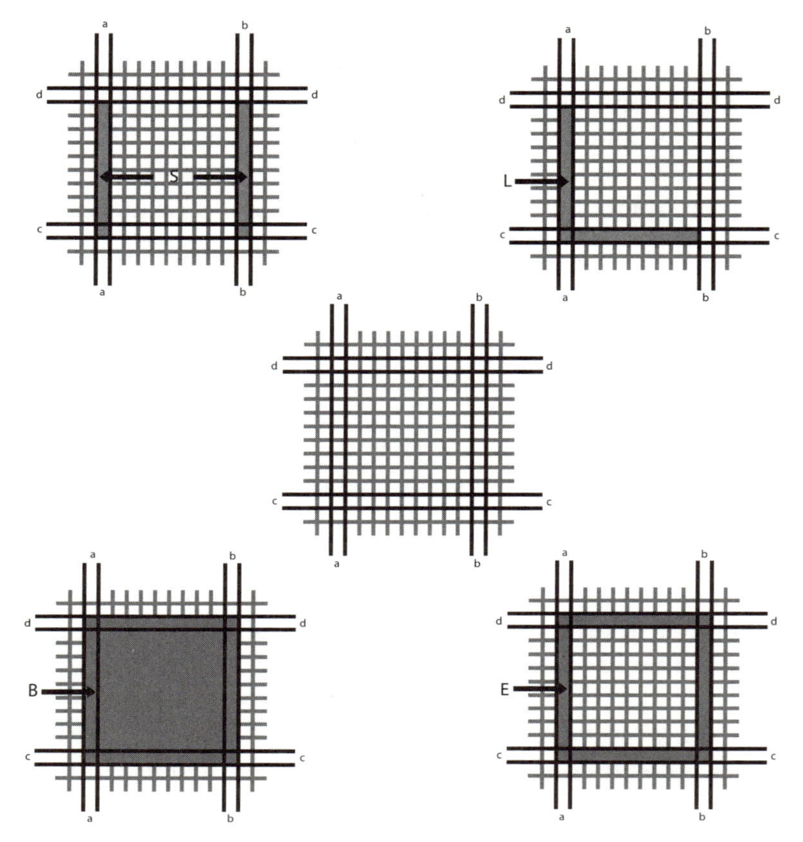

Figure 8.9 The $k \times k$ sub-board B.

shown in Figure 8.9. Each of the i queens above or below B lying between a and b then controls exactly 6 squares in E. Similarly, each of the i queens to the right or left of B lying between c and d controls exactly 6 squares in E. All of the rest of the queens—and there are $\frac{1}{2}(n-1) - 2i$ of them—are inside the square sub-board B, and each of these queens controls exactly 8 squares in E.

So, we can conclude that at most $6i + 6i + 8(\frac{1}{2}(n-1) - 2i)$ squares in E are controlled. This simplifies to $4(n - i - 1) = 4(k-1)$ squares in E that are controlled, but this is also exactly the *total* number of squares in E. This means that each square in E is controlled by a unique queen.

Note that the corner squares of B are necessarily covered from within B; hence, the two main diagonals of B are occupied, which means that there is guaranteed to be at least one queen in B. Thus, $\frac{1}{2}(n-1) \geqslant 2i+1$. Since $i = n - k$, it follows that $k \geqslant \frac{3}{4}(n+1)$, and this tells us quite a bit about the size of the square sub-board B.

Let L be the 'L'-shaped set of squares along the left side and the bottom of E that are below row d and to the left of column b. So, the set L contains $2k-3$ squares. Now, each of the $\frac{1}{2}(n-1)$ queens controls exactly one square of L along a positive diagonal. Moreover, each of the $2i = 2(n-k)$ queens outside of B controls exactly one square of L orthogonally, while each of the queens inside of B controls exactly two squares of L orthogonally. Finally, there are a few queens, say j of them, that are below the main negative diagonal of B, as extended to the entire chessboard, that control exactly two squares of L along a negative diagonal.

Thus, since each of the $2k-3$ squares in L is controlled by just one queen, we have

$$\frac{1}{2}(n-1) + 2(n-k) + 2(\tfrac{1}{2}(n-1) - 2(n-k)) + 2j = 2k-3,$$

and this reduces to $n = 4j + 3$, thus showing that $n \equiv 3 \bmod 4$, as desired.

Finally, we can show that k must be an odd number. First, we call a square (x, y) *even* or *odd* according to whether $x+y$ is even or odd. Suppose that k is even. Then E has an equal number of odd and even squares. Also, any queen controlling squares of E along a row or column does so at one even and one odd square. Thus, the squares of E that are controlled diagonally must have an equal number of odd and even squares, and so the number of 'odd' queens equals the number of 'even' queens. Hence, the number of queens is itself even. But, when $n \equiv 3 \bmod 4$, $\frac{1}{2}(n-1)$ is odd, a contradiction, and so k is odd. This completes the proof of the theorem. □

Now, after all that work, we can continue with the next entry, $n = 7$, in our list of queens domination numbers as a corollary

Figure 8.10 $\gamma(Q_{7\times 7}) = 4$.

to some of the facts that emerged during the proof of Theorem 8.3.

Corollary 1 $\gamma(Q_{7\times 7}) = 4$.

Proof. Suppose that the 7×7 chessboard can be dominated by just 3 queens. In other words, since $\frac{1}{2}(7-1) = 3$, we are supposing that the situation of Theorem 8.3 holds. Then, as we saw in the proof of that theorem, $k \geqslant \frac{3}{4}(n+1) = 6$, but k must also be odd, which means that $k = 7$, and so, in this case, the square sub-board B is in fact the entire chessboard, and each edge square of the 7×7 board is controlled by a single queen. If the corner squares are controlled by a single queen placed in the center, then there is no place to put a second queen so that it doesn't also control one of the 8 edge squares already controlled by the first queen. The same sort of thing happens if we begin by placing the first queen at $(2, 2)$ to cover two corner squares. It is impossible to place a second queen. A final choice for placing the first queen is $(3, 3)$ and this once again makes it impossible to place a second queen. Thus, $\gamma(Q_{7\times 7}) > 3$. So, $\gamma(Q_{7\times 7}) \geqslant 4$, and with this as a lower bound, the covering shown in Figure 8.10 demonstrates that $\gamma(Q_{7\times 7}) = 4$. □

For $n = 8$, Spencer's lower bound is given by $\gamma(Q_{8\times 8}) \geqslant \frac{1}{2}(8-1) = \frac{7}{2}$, and so $\gamma(Q_{8\times 8}) \geqslant 4$, which isn't particularly helpful since it has long been known that the true value is $\gamma(Q_{8\times 8}) = 5$, as was claimed by W. W. Rouse Ball, without proof.

The next value of n to be considered, $n = 9$, is of the form $4k + 1$, which is to say it is *not* of the form $4k + 3$, and so, by Theorem 8.3, Spencer's lower bound of $\frac{1}{2}(n-1)$ could not be achieved here. Therefore, for n of the form $4k + 1$ the lower

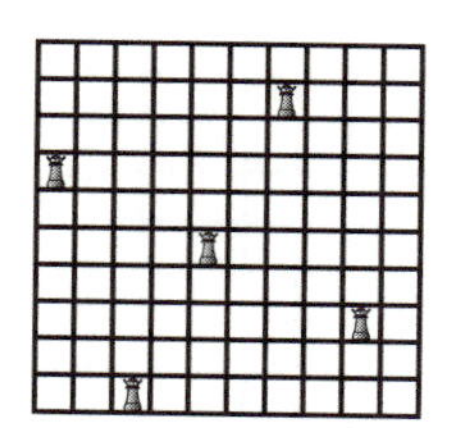
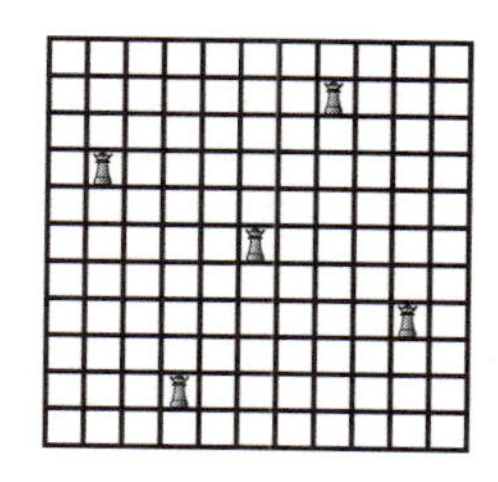
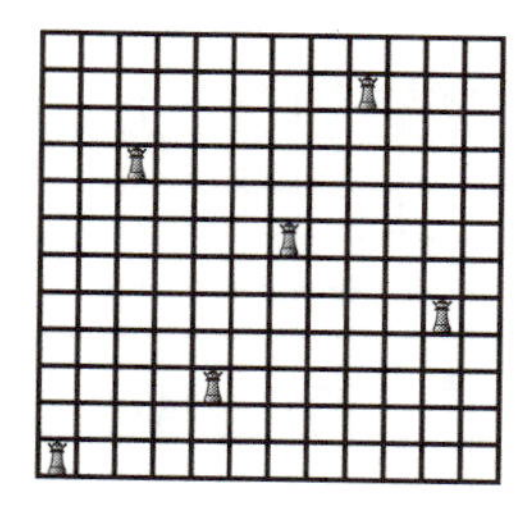

Figure 8.11 $\gamma(Q_{10\times10}) = \gamma(Q_{11\times11}) = 5$ and $\gamma(Q_{12\times12}) = 6$.

bound needs revising, and the new lower bound will become $\frac{1}{2}(n-1)+1 = \frac{1}{2}(n+1)$.

Corollary 2 *For* $n = 4k+1$, $\gamma(Q_{n\times n}) \geqslant \frac{1}{2}(n+1) = 2k+1$.

This is a particularly robust corollary and, as we shall see, one that frequently provides exact values for the queens domination number. For example, for $n = 9$, Corollary 2 tells us that $\gamma(Q_{9\times9}) \geqslant 5$, and we know from Figure 8.4 that 5 queens can cover a 9×9 board, and so $\gamma(Q_{9\times9}) = 5$.

For $n = 10$, we are back to Spencer's lower bound of $\frac{1}{2}(n-1)$ from Theorem 8.2 for an even value of n, which gives us 5 as a lower bound. We know that the arrangement of 5 queens from Figures 8.4 and 8.5 isn't going to work for the 10×10 board. But the strikingly similar arrangement of 5 queens shown in Figure 8.11 does work to cover the 10×10 board, and therefore yields $\gamma(Q_{10\times10}) = 5$.

The last known odd value of n for which Spencer's lower bound is achieved is $n = 11$. Here the lower bound is 5, and the winning arrangement is shown in Figure 8.11. Thus, $\gamma(Q_{11\times11}) = 5$. Of course, this arrangement, which is the same one used for the 10×10 board, could also be used on the 9×9 board.

Spencer's lower bound is also achieved, however, for the next even value of n, that is, for $n = 12$. Here the lower bound is $\frac{1}{2}(12-1) = \frac{11}{2}$, which gives us 6 as a lower bound, and it is then easy to use the same arrangement of 5 queens from the 11×11 board, and simply add a 6th queen in a corner, as shown in Figure 8.11, in order to cover the 12×12 board. Therefore, $\gamma(Q_{12\times12}) = 6$.

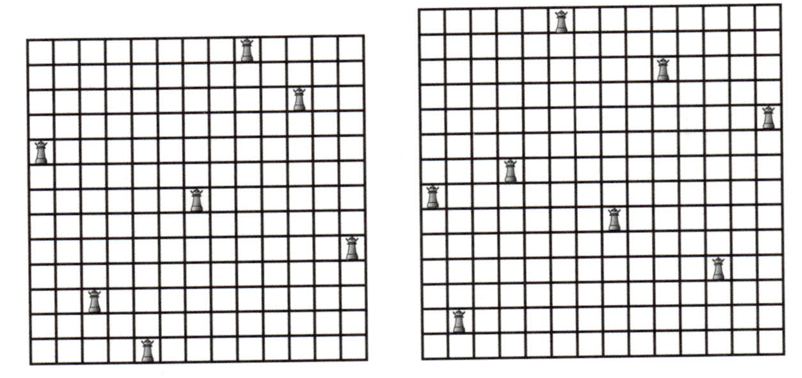

Figure 8.12 $\gamma(Q_{13\times13}) = 7$ and $\gamma(Q_{14\times14}) \leqslant 8$.

For $n = 13$, we can again use Corollary 2, which tells us that $\gamma(Q_{13\times13}) \geqslant 7$, and in Figure 8.12 we show a covering of the 13×13 board with 7 queens given by Burger, Mynhardt, and Cockayne [4], and so $\gamma(Q_{13\times13}) = 7$.

We now come to our first value of n for which the queens domination number is not known! For $n = 14$, Spencer's lower bound is $\frac{1}{2}(14-1) = \frac{13}{2}$, which tells us that $\gamma(Q_{14\times14}) \geqslant 7$. Trying to cover the 14×14 board with 7 queens, we can place a queen at $(2, 2)$, and then begin a series of 'super' knight's moves on a torus of the form $(2, 6)$—that is, 'over 2 and up 6'—and place a total of 7 queens on the board before this repeating 'super' knight's move returns us once again to the original square $(2, 2)$. These 7 queens then cover the entire 14×14 chessboard with the exception of just the two squares symmetrically positioned at $(1, 7)$ and $(7, 1)$. Placing an 8th queen at either of these two squares then covers the entire 14×14 board, as shown in Figure 8.12. So, $\gamma(Q_{14\times14})$ is either 7 or 8. This situation is somewhat reminiscent of what we saw in Figure 1.12 in Chapter 1 with 4 queens placed on the 8×8 board also missing just two squares, and I suspect that the domination number here is 8.

Things now begin to get worse for us. For $n = 15$, the lower bound from Theorem 8.2 is 7. For $n = 16$, the lower bound is 8. For $n = 17$, we can again use Corollary 2 which tells us that the lower bound for the 17×17 board is 9. The arrange-

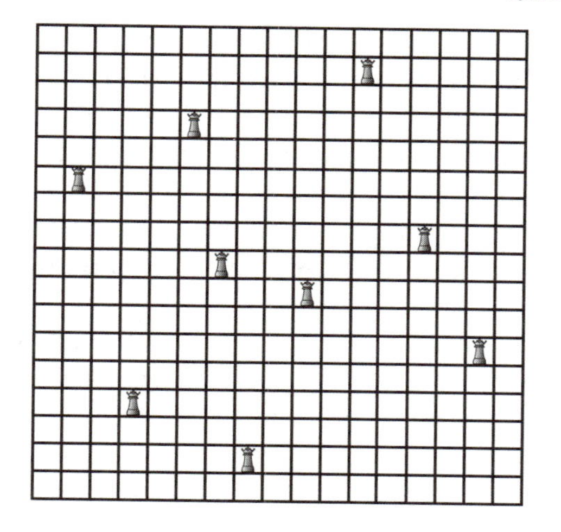

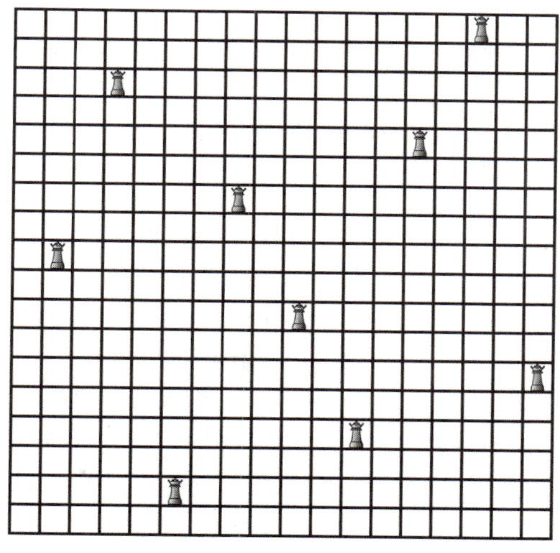

Figure 8.13 $\gamma(Q_{15\times15}) \leqslant 9$, $\gamma(Q_{16\times16}) \leqslant 9$, $\gamma(Q_{17\times17}) = 9$; and $\gamma(Q_{18\times18}) = 9$, $\gamma(Q_{19\times19}) \leqslant 10$.

ment of 9 queens shown on the top in Figure 8.13 covers the 17×17 board, and so $\gamma(Q_{17\times17}) = 9$. Since this same arrangement can obviously be used to cover both the 15×15 board and the 16×16 board, we can at least conclude that $\gamma(Q_{15\times15}) = 7$, 8, or 9; and that $\gamma(Q_{16\times16}) = 8$ or 9.

For $n = 18$, however, things brighten again, at least momentarily. Since n is even, the lower bound from Theorem 8.2 is 9, and an arrangement, shown on the bottom in Figure 8.13, with 9 queens covering the 18×18 board was found by McRae. Thus, $\gamma(Q_{18\times18}) = 9$.

For $n = 19$, the lower bound is 9, and we can use McRae's arrangement of 9 queens for the 18×18 board in Figure 8.13, and add a 10th queen in a corner formed by adding a new row and a new column, and thus cover a 19×19 board. Therefore, $\gamma(Q_{19\times19}) = 9$ or 10.

For $n = 20$, the lower bound is 10. But then, for $n = 21$, we can use Corollary 2, which tells us that the lower bound for the 21×21 board is 11. The arrangement of 11 queens shown on the top in Figure 8.14 covers the 21×21 board, and so $\gamma(Q_{21\times21}) = 11$. Since this same arrangement can obviously be used to cover the 20×20 board, we can also conclude that $\gamma(Q_{20\times20}) = 10$ or 11.

For $n = 22$, the lower bound is 11, and we can use the arrangement of 11 queens for the 21×21 board in Figure 8.14, and then add a 12th queen in a corner formed by adding a new row and a new column to cover a 22×22 board. Thus, $\gamma(Q_{22\times22}) = 11$ or 12.

For $n = 23$, the lower bound is 11. For $n = 24$, the lower bound is 12. For $n = 25$, however, we can use Corollary 2, which tells us that the lower bound is 13. The arrangement of 13 queens shown on the bottom in Figure 8.14 covers the 25×25 board, and so $\gamma(Q_{25\times25}) = 13$. Since this same arrangement can obviously be used to cover both the 23×23 board and the 24×24 board, we can at least conclude that $\gamma(Q_{23\times23}) = 11$, 12, or 13, and that $\gamma(Q_{24\times24}) = 12$ or 13.

I'll conclude this chapter on queens domination with a problem on rectangular chessboards. David Fisher has done some interesting work on the queens domination problem for rectangular chessboards and, in particular, has produced a set of computer-generated values for boards where the domination number is less than 10.

Problem 8.4 It is obvious that a rectangular chessboard with only four or fewer rows can be dominated by four queens, no matter how many columns it has. However, for chessboards having more than four rows, the largest rectangular boards that

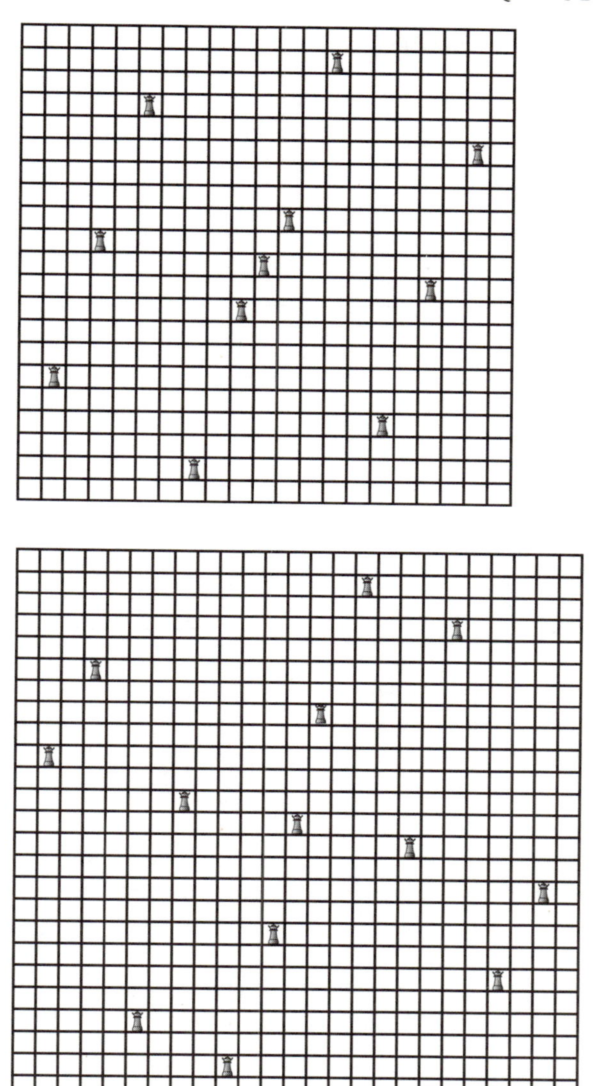

Figure 8.14 $\gamma(Q_{20\times20}) \leqslant 11$, $\gamma(Q_{21\times21}) = 11$, $\gamma(Q_{22\times22}) \leqslant 12$; and $\gamma(Q_{23\times23}) \leqslant 13$, $\gamma(Q_{24\times24}) \leqslant 13$, $\gamma(Q_{25\times25}) = 13$.

can be dominated by four queens are the 5×12 chessboard and the 6×10 chessboard. Find arrangements of four queens that dominate each of these rectangular chessboards.

SOLUTIONS TO PROBLEMS

Solution 8.1 Solutions for the main diagonal, the sub-diagonal, and a central column are shown in Figure 8.15.

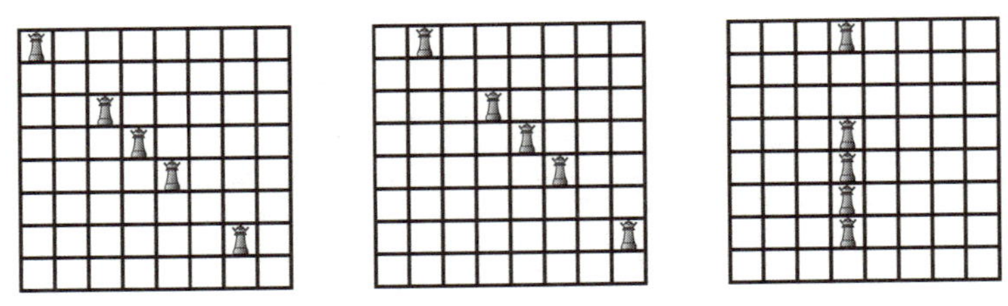

Figure 8.15 The 8×8 chessboard dominated by five queens in a row.

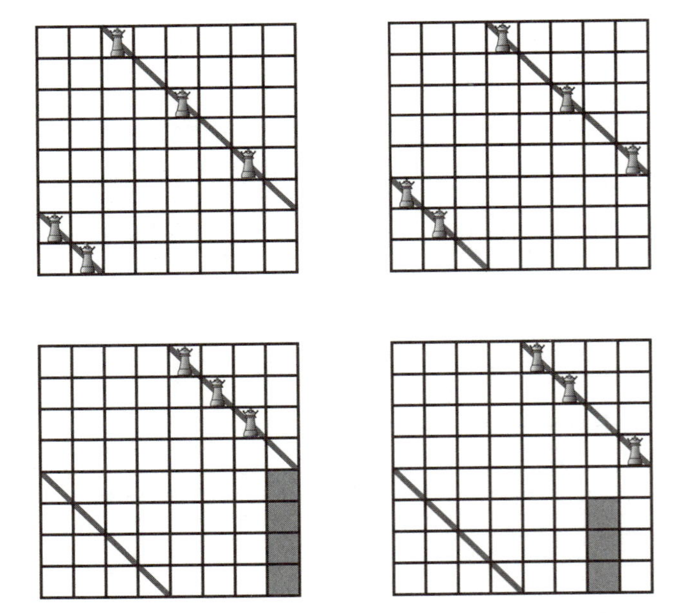

Figure 8.16 Covering the 8×8 board along *toroidal* diagonals.

This same diagonal pattern from the first two chessboards in Figure 8.15 can be repeated to provide solutions for two more of the toroidal diagonals, as shown on the top in Figure 8.16. However, the remaining toroidal diagonal—the last one shown in Figure 8.2—has no solution with five queens. Here is a proof. Since this last diagonal splits evenly on the chessboard, three or more queens must lie along one of its two halves. Clearly, if four queens are placed in one of the two halves, a single queen cannot be placed in the other half and cover the remaining eight uncovered squares. On the other hand, if three queens are placed in one of the two halves, then, because of symmetry, this can be done in only two ways, and these are both shown on the bottom in Figure 8.16. In each of these two cases, the shaded squares can't be covered by the remaining two queens no matter where they are placed along the remaining half of the diagonal.

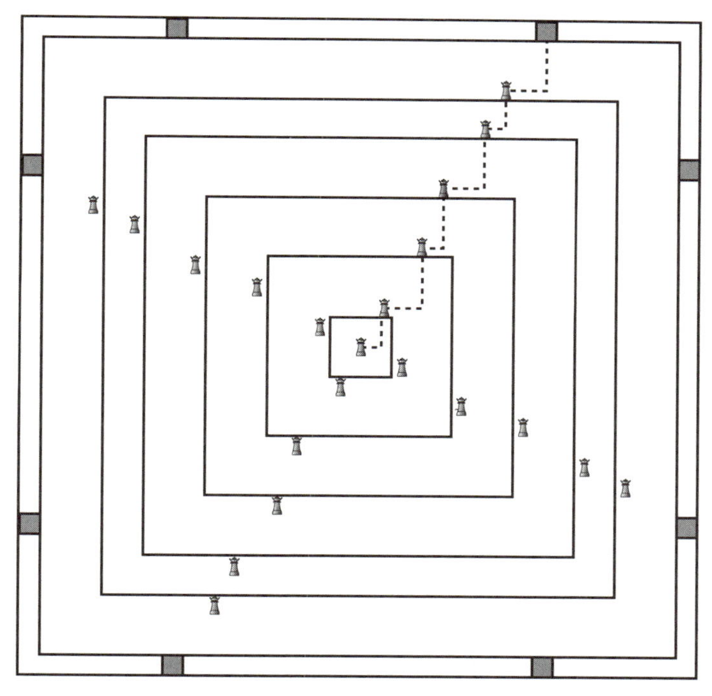

Figure 8.17 The Spencer-Cockayne construction.

Solution 8.2 The surprise is that you don't always gain a ring of three rows at each stage of the construction. We began with a 3×3 square for one queen, and then got a 9×9 square for five queens, and a 15×15 square for nine queens. The next iteration continues this pattern yielding a 21×21 square for thirteen queens. Then comes the surprise: the next four queens only control a 25×25 square instead of the 27×27 square you might have expected. Figure 8.17 shows the result of one additional iteration. I refer you to Cockayne [6] for a discussion of exactly how the size of the squares are growing in this construction.

I would also like to suggest an alternate way to view this construction. Begin with the very first queen, then the next four queens are each placed a knight's move away, a (1,2)-move, if you will, each move rotated 90° from the previous one. Similarly, you can think of the second stage as a knight's (2,3)-move from the previous four queens, and the next stage as a (1,3)-move, and the next as a (2,3)-move, and then back to a (1,2)-move—the surprise!—and then a (2,3)-move again. As this continues, it is always the (1,2)-move that results in a gain of only a ring of two additional rows.

Solution 8.3 For an $n \times n$ board, we can place one queen in the center of the 3×3 upper left-hand corner, and then place the remaining $n - 3$ queens along the main diagonal to cover the rest of the board, as shown in Figure 8.18.

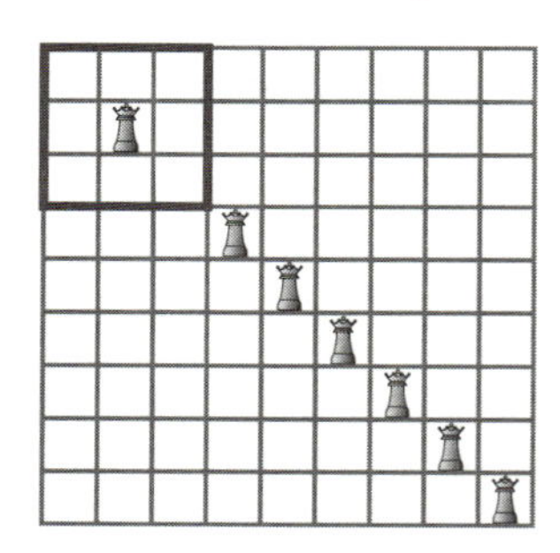

Figure 8.18 $\gamma(Q_{n \times n}) \leqslant n - 2$.

Solution 8.4 These arrangements are shown in Figure 8.19.

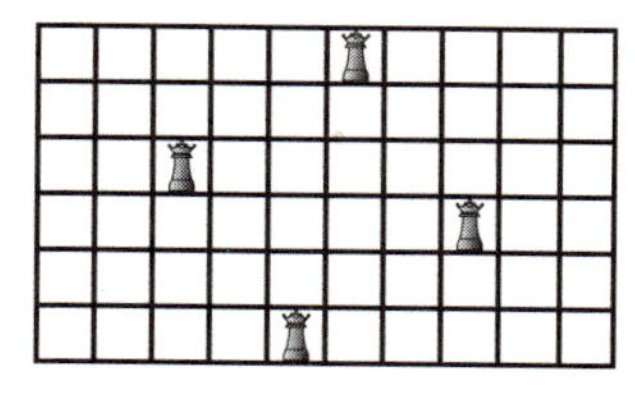

Figure 8.19 $\gamma(Q_{5\times12}) = \gamma(Q_{6\times10}) = 4$.

Domination on Other Surfaces

We closed the last chapter with the observation that among chessboards with more than four rows, the 5×12 chessboard and the 6×10 chessboard are the largest boards that can be dominated by four queens. In particular, the four queens in Figure 1.12 in Chapter 1 failed to cover the 8×8 chessboard shown there. However, as we noted earlier, they only missed two squares on that board, and if we were to think of this chessboard as being on a torus, then those two squares *would* be covered. In fact, it would even be sufficient for us to place this chessboard on a cylinder in order for these four queens to dominate the entire 8×8 board. This suggests that it might be interesting to take another look at the domination problem for each of the chess pieces, and consider what happens on various surfaces such as the torus and the Klein bottle for the queen, the knight, the rook, the bishop, and the king. We'll start with a problem.

Problem 9.1 How many of each chess piece does it take to cover the $2 \times 2 \times 2$ cube? That is, how many knights? How many bishops? How many rooks? How many queens? How many kings?

The Queens Graph on the Torus

We will denote the queens graph for an $n \times n$ chessboard on the torus by $Q_{n\times n}^{\text{tor}}$, and so $\gamma(Q_{n\times n}^{\text{tor}})$ is the domination number for this graph. Obviously, then, for any n, $\gamma(Q_{n\times n}^{\text{tor}}) \leqslant \gamma(Q_{n\times n})$. A comparison of the first ten domination numbers for the queens graph and the toroidal queens graph is given in Table 9.1.

Table 9.1 Comparing queens domination numbers.

n	1	2	3	4	5	6	7	8	9	10
$\gamma(Q_{n\times n})$	1	1	1	2	3	3	4	5	5	5
$\gamma(Q^{\text{tor}}_{n\times n})$	1	1	1	2	3	3	4	4	5	5

Among these first ten values, the toroidal domination number turns out to be actually *less than* the domination number only for the one board we happened to have discussed already, that is, for the 8×8 chessboard. The others are all the same.

The Knights Graph on the Torus

We can do the same thing for the knights graph on the torus. A comparison of the first eight domination numbers for the knights graph and the toroidal knights graph is given in Table 9.2.

Table 9.2 Comparing knights domination numbers.

n	1	2	3	4	5	6	7	8
$\gamma(N_{n\times n})$	1	4	4	4	5	8	10	12
$\gamma(N^{\text{tor}}_{n\times n})$	1	2	3	4	5	6	9	8

You might have noticed that something extraordinary just happened here. In going from the 7×7 chessboard on the torus to the 8×8 chessboard on the torus, the number of knights needed to cover the larger of the two chessboards actually dropped! It dropped from nine to eight. This is very unexpected and completely counterintuitive. In fact, one of the long-time outstanding unsolved problems for the queens graph is whether or not it is always true that $\gamma(Q_{n\times n}) \leqslant \gamma(Q_{(n+1)\times(n+1)})$; in other words, is the queens domination number monotonic? So, the discovery that the *toroidal* knights domination number is not monotonic is quite surprising, to say the least.

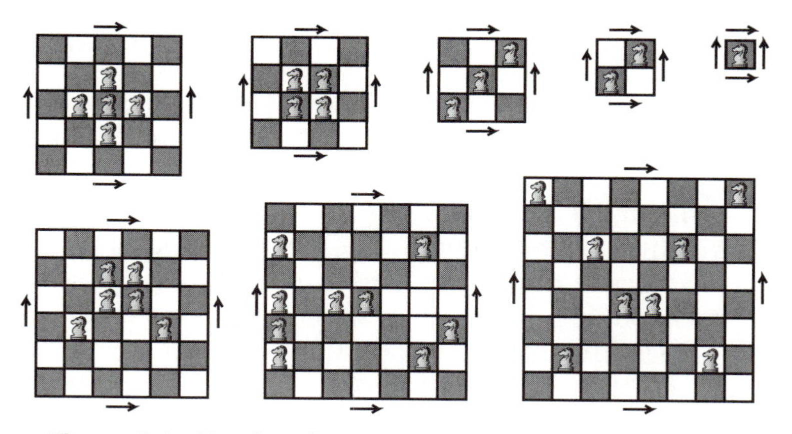

Figure 9.1 Knights dominating chessboards on a torus.

A knights cover of each of the toroidal chessboards up to the size 8×8 board is shown in Figure 9.1. For $n = 1$, of course, only one knight is required. For $n = 2$, four knights are required on a regular 2×2 chessboard, one for each square, but only two knights are required on a torus. For $n = 3$, four knights are again required on a regular 3×3 chessboard, but three will do on a torus. And, it is easy to see by inspection that two knights can't cover a 3×3 board on a torus: there are effectively only three different squares on which to place a knight, and in each case a knight covers a total of only five squares including the square it occupies.

For $n = 4$, on a regular 4×4 chessboard the best you can do is put four knights in the four center squares. On a torus, any single knight covers exactly five squares, so three knights won't be sufficient to cover a 4×4 board, and you still need four knights, just as on a regular board.

For $n = 5$, you need five knights on a regular 5×5 chessboard, and placing them in the center works. Surprisingly, you still need five knights on a torus. The best you can do with four knights is to cover nine squares with the first knight, seven squares with the second knight, five more squares with the third, and three more with the fourth, leaving one square still uncovered.

For $n = 6$, a regular 6×6 chessboard needs eight knights, but on a torus you can do it with six knights. Note that since n is even, a knight on a white square only attacks black squares, and a knight on a black square only attacks white squares. In the language of graph theory, then, we call such a graph *bipartite*, since each edge has one black vertex and one white vertex. A single knight on a white square attacks eight black squares, and the best you can do with two knights placed on white squares is to attack fourteen black squares. Hence, on a torus, you need to use at least three knights placed on white squares to cover the eighteen black squares of the 6×6 board. You might want to check that using only two knights placed on white squares to cover fourteen black squares, and then placing four knights on the four uncovered black squares doesn't quite work. In fact, it turns out that the only way to cover the eighteen black squares is with three knights in a row diagonally on white squares. So, to cover the 6×6 board on a torus, you need six knights, three for the black squares and three for the white squares, but the three knights in a row on white squares and the three knights in a row on black squares can be placed anywhere.

Note the amusing fact that up to this point we have concluded that, for $n \leqslant 8$, $\gamma(N^{\text{tor}}_{n \times n}) = n$, except for $n = 7$.

For $n = 7$, ten knights are required for a regular 7×7 chessboard. The arrangement of nine knights shown for the 7×7 board in Figure 9.1 shows that on a torus $\gamma(N^{\text{tor}}_{7 \times 7}) \leqslant 9$. Moreover, a student, Sarah Eisen, has shown by a computer search that the 7×7 board cannot be covered on a torus by fewer than nine knights, and so $\gamma(N^{\text{tor}}_{7 \times 7}) = 9$.

For $n = 8$, things get more interesting. A regular 8×8 chessboard needs twelve knights. But, on a torus, you can cover the 8×8 board with just eight knights, and do it with near-perfect efficiency, in the sense that each unoccupied square is controlled by only one knight. So, the four knights placed on white squares in Figure 9.1 cover the 32 black squares, and the four knights placed on black squares cover the 32 white squares. It is this remarkable efficiency that not only creates the dramatic drop from twelve knights for the regular chessboard down to eight knights for the toroidal chessboard, but also results in

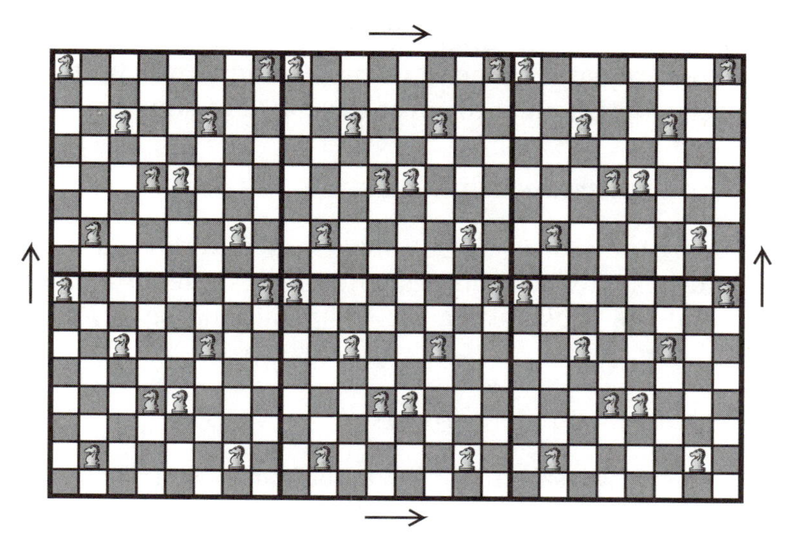

Figure 9.2 A near-perfect knights cover of the 16×24 chessboard on a torus.

the astonishingly bizarre fact that fewer knights are needed for the 8×8 chessboard than are needed for the smaller 7×7 chessboard on a torus.

Moreover, on a torus, we can piece together copies of this 8×8 chessboard to find near-perfect—and almost undoubtedly optimal—coverings for any $m \times n$ chessboard where m and n are multiples of 8. An example is shown in Figure 9.2 for a 16×24 chessboard. Since these boards can be covered so efficiently, it seems likely that the *toroidal* knights domination number, $\gamma(N^{\text{tor}}_{n \times n})$, will continue to exhibit non-monotonic behavior every time it reaches a value of n that is a multiple of 8.

We have seen that the toroidal knights domination number, $\gamma(N^{\text{tor}}_{n \times n})$, fails to be monotonic, but it may nonetheless possess a different, and still rather interesting, property: namely, that the $n \times n$ chessboard on the torus has a *unique* domination number for each value of n.

Problem 9.2 Show that at least twelve knights are required to cover a regular 8×8 chessboard by finding twelve squares having the property that no matter where you place a knight on the chessboard the knight can never control or occupy any two of these squares at the same time. In other words, it would take twelve knights just to cover these twelve squares.

THE ROOKS GRAPH ON THE TORUS

Since being on a torus doesn't make any difference at all to a rook in terms of which squares it dominates, the *toroidal rooks graph* and the *rooks graph* are identical, and so $\gamma(R^{\text{tor}}_{n\times n}) = \gamma(R_{n\times n}) = n$.

THE BISHOPS GRAPH ON A TORUS

In Chapter 7 we saw that, for a regular $n \times n$ chessboard, $\gamma(B_{n\times n}) = n$. The argument presented there, to be sure, was somewhat detailed since it involved separate arguments for the black squares and the white squares, as well as a transformation of the bishops into rooks. On a torus, however, it is much easier for us to argue that $\gamma(B^{\text{tor}}_{n\times n}) = n$, since there are now n diagonals in each direction, and, with fewer than n bishops, you would have an unoccupied diagonal in each direction, and so the intersection of these diagonals would be an uncovered square.

I'll mention in passing that when n is odd the toroidal bishops graph, $B^{\text{tor}}_{n\times n}$, is equivalent to the $n \times n$ rooks graph, $R_{n\times n}$. In the language of graph theory, we call such graphs *isomorphic*. But, when n is even, these graphs are actually quite different; for example, the toroidal bishops graph is in this case a disconnected graph with two identical components, whereas the rooks graph is a connected graph. Another difference, when n is even, is that, in the toroidal bishops graph, a positive diagonal and a negative diagonal intersect twice if they are in the same component of the graph; whereas, of course, in the rooks graph a horizontal row and a vertical row only intersect once.

THE KINGS GRAPH ON A TORUS

It will probably now come as a surprise to you that the most interesting thing that happens to domination on a torus hap-

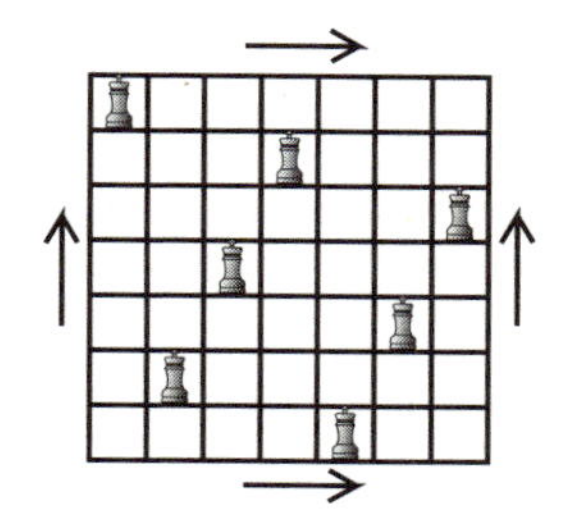

Figure 9.3 Kings dominating a regular 7×7 chessboard and a toroidal 7×7 chessboard.

pens for the one chess piece that covers regular chessboards in the most rigid, and completely boring, grid-like fashion of all: namely, the king. You may recall from Chapter 7 that the regular 7×7 chessboard requires nine kings to be covered, and that this is just as many kings as the regular 9×9 chessboard requires. In other words, on a regular chessboard you can't do any better than use the simple, and boring, lattice pattern for these nine kings shown on the left in Figure 9.3. But, on a torus, it turns out you can cover the 7×7 chessboard with only seven kings by using the pattern shown on the right in Figure 9.3, where we start by placing a king in the upper left-hand corner, and then going over three squares and down one square in order to place each of the next six kings. At this point, you might want to check for yourself that each of the 49 squares on this toroidal chessboard are in fact covered by one of the seven kings.

In the following problem we see that a pattern similar to the one used in Figure 9.3 for the 7×7 board can be used on a torus to cover boards of size 16×16, 25×25, 34×34, and so on. This pattern uses 4 fewer kings on the 16×16 board than the standard lattice pattern does, 6 fewer on the 25×25 board, 8 fewer on the 34×34 board, and so on.

Problem 9.3 Show that, for any n of the form $n = 9k + 7$, the $n \times n$ chessboard can be covered on a torus with $n(k + 1)$ kings, using $k + 1$ kings in each row. Thus, on a torus, you need $2k + 2$ fewer kings than you need on a regular chessboard.

Now, of course, when n is a multiple of 3, even on a torus, you can't do better than use the standard lattice pattern for kings. In other words, for $n = 3k$, $\gamma(K_{n\times n}^{\text{tor}}) = \gamma(K_{n\times n}) = (\frac{1}{3}n)^2$. A student, Chris Ricci, was the first to realize that for any other value of n, just as we saw above, you can always do better on a torus than on a regular chessboard. Let's see precisely how much better you can hope to do by first looking for a lower bound for the kings domination number on a torus.

As it happens, there is an exceedingly easy counting argument at hand that will give us a lower bound for the toroidal kings domination number for the $n \times n$ chessboard. The key observation is that any given row has n squares that must be covered by kings. Therefore, there must be at least $\lceil \frac{1}{3}n \rceil$ kings that have to be involved in covering these n squares. Here, '$\lceil x \rceil$' represents the *ceiling function*, which behaves just like the floor function except that you round *up* instead of *down*. Where can these $\lceil \frac{1}{3}n \rceil$ kings be? They must either lie in the given row itself, in the row immediately above the given row, or in the row immediately below the given row. Thus, we see that each horizontal band of three rows must contain at least $\lceil \frac{1}{3}n \rceil$ kings. Since there are exactly n such bands—each row has a 3-row band centered on it, after all—and since each king gets counted three times in this way, we conclude that there must be at least $\lceil \frac{1}{3}n \lceil \frac{1}{3}n \rceil \rceil$ kings in any cover. This is our lower bound.

Note that, when n is divisible by 3, this lower bound, not surprisingly, reduces to $(\frac{1}{3}n)^2$. Thus, in this particular case, the lower bound actually gives us the exact toroidal kings domination number. Note, also, that the lower bound gives us the exact toroidal kings domination number in the case where $n = 9k + 7$, because in that case, we have a lower bound given by

$$\left\lceil \frac{n\lceil \frac{1}{3}n \rceil}{3} \right\rceil = \left\lceil \frac{n\lceil \frac{1}{3}(9k+7) \rceil}{3} \right\rceil = \lceil \tfrac{1}{3}n(3k+3) \rceil$$
$$= \lceil n(k+1) \rceil = n(k+1),$$

and, as we saw in Problem 9.3, the $n \times n$ chessboard could actually be covered with $n(k+1)$ kings on a torus.

Nonetheless, it is still a pleasant surprise for us to discover that this lower bound *always* gives us the toroidal kings domination number, especially given how easy this lower bound was to find. Thus, we have the following:

$$\gamma(K_{n\times n}^{\text{tor}}) = \text{the toroidal kings domination number} = \left\lceil \frac{n\lceil \frac{1}{3}n \rceil}{3} \right\rceil.$$

If we prefer, this can be broken down into cases and written as

$$\gamma(K_{n\times n}^{\text{tor}}) = \begin{cases} \frac{1}{9}n^2 & \text{for } n = 3k; \\ \lceil \frac{1}{9}(n^2 + 2n) \rceil & \text{for } n = 3k + 1; \\ \lceil \frac{1}{9}(n^2 + n) \rceil & \text{for } n = 3k + 2. \end{cases}$$

The examples considered in Figure 9.3 and Problem 9.3 give the impression that on a torus the kings domination number is always achieved by having the same number of kings in each row. But, of course, the above formulas for $\gamma(K_{n\times n}^{\text{tor}})$ do not always yield numbers of kings that are evenly divisible by n. We can illustrate the general procedure for placing the appropriate number of kings on an $n \times n$ chessboard by considering the 14×14 board. First, note that computing the lower bound gives us

$$\left\lceil \frac{14\lceil \frac{14}{3} \rceil}{3} \right\rceil = \left\lceil \frac{14(5)}{3} \right\rceil = \left\lceil \frac{70}{3} \right\rceil = 24$$

kings. We don't actually need to know this number ahead of time in order to carry out the procedure, but it will be comforting nonetheless to see that we end up with the right number of kings in the end.

The next thing to do is to decide how many kings to put in each of the fourteen rows. Since each row has fourteen squares, we need five kings to cover these squares, and so any three consecutive rows need to contain at least five kings. We begin by placing two kings in the first row, two kings in the second row, and one king in the third row. The pattern then forces itself upon us: we must place two in the fourth row, and so on. The only time we can place one king in a row instead of two kings is when we have placed two kings in each of the two previous rows. The pattern, then, for the

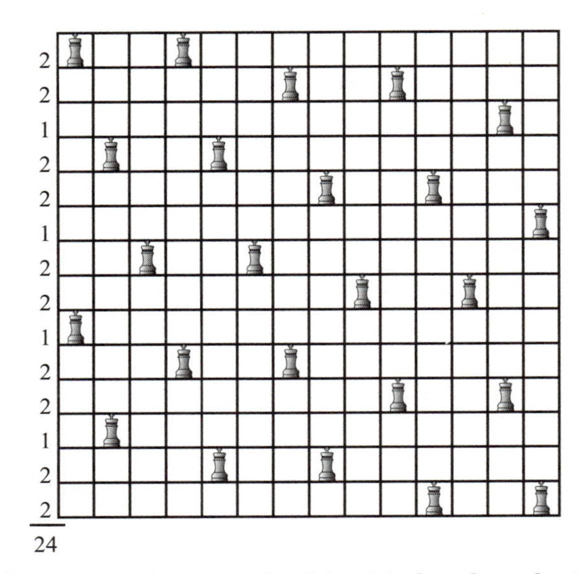

Figure 9.4 Covering the 14×14 chessboard with 24 kings on a torus.

fourteen rows is: $2, 2, 1, 2, 2, 1, 2, 2, 1, 2, 2, 1, 2, 2$. By the way, this is a total of 24 kings. In general, when carrying out this procedure, there are only three patterns that ever need be used: namely, the one we just used, $x, x, x-1, x, x, x-1, \ldots$; $y, y-1, y-1, y, y-1, y-1, \ldots$; and the one used in Figure 9.3 and Problem 9.3, $z, z, z, z, z, z, \ldots$.

Now, the actual placement of these 24 kings is completely straightforward. Begin with the first king in the upper left-hand corner, and then always place the next king three squares to the right, and, either in the same row or in the next row down, whichever is appropriate based on how many kings the row you are working in is supposed to have. This whole procedure is illustrated in Figure 9.4 for the 14×14 board.

Note that, as we worked our way down the board placing kings in this example, each row we came to was guaranteed to be covered precisely because we previously arranged to have enough kings available at each stage in the band of three rows surrounding a given row. Note, also, that the two kings in the bottom row cover the two squares near the end of the top row that were previously uncovered, and that the first two kings at the top do the same thing for two squares in the bottom row. The algebraic details of showing that we can always count on such a fortuitous occurrence can be found in [34], where, in fact, the following more general result about kings dominating rectangular toroidal chessboards is presented.

Theorem 9.1 (Watkins and Ricci) *The toroidal kings domination number for a rectangular $m \times n$ chessboard is given by*

$$\gamma(K_{m\times n}^{\text{tor}}) = \max\left\{\left\lceil \frac{m\lceil \frac{1}{3}n\rceil}{3}\right\rceil, \left\lceil \frac{n\lceil \frac{1}{3}m\rceil}{3}\right\rceil\right\}.$$

In particular, then,

$$\gamma(K_{n\times n}^{\text{tor}}) = \left\lceil \frac{n\lceil \frac{1}{3}n\rceil}{3}\right\rceil.$$

Essentially, this theorem says that for a rectangular chessboard on a torus there are two lower bounds to be computed: one bound is computed by considering rows, just as we did for square boards; and the other is computed by considering columns in exactly the same way. Since these two computations may yield different numbers, we simply take whichever number turns out to be the greater as our lower bound. Problem 9.4 explores how to go about achieving this lower bound.

Problem 9.4 Show how to dominate the 7×10 and the 8×10 chessboards with the minimum number of kings on a torus.

The Kings Graph on a Klein Bottle

The situation for kings domination on the Klein bottle is somewhat easier to analyze than it was on a torus. In Figure 9.5, we show how to cover the 7×7 board on the Klein bottle with eight kings. Note that this is not quite as good as we could do on the torus, but it is still better than we could do on a regular chessboard. The basic idea is that, since the right-hand edge of the band of three rows at the top connects with the left-hand edge of the band of three rows at the bottom, we can cover

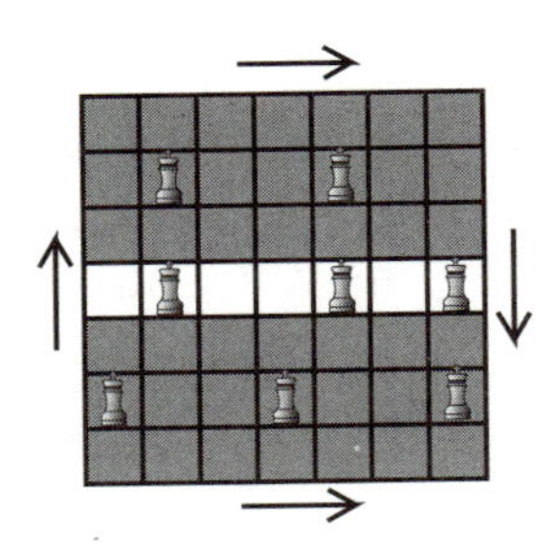

Figure 9.5 Kings dominating a 7×7 chessboard on a Klein bottle.

this shaded 3×14 band with just five kings placed in the middle row of the band. We can then cover the remaining seven squares of the middle row with three kings.

In the following problem we see that a pattern similar to the one just used for the 7×7 board can be used on a Klein bottle to cover chessboards of size 13×13, 19×19, 25×25, and so on. This pattern uses 2 fewer kings on the 13×13 board than the standard lattice pattern does, 3 fewer on the 19×19 board, 4 fewer on the 25×25 board, and so on.

Problem 9.5 Show that for any n of the form $n = 6k + 1$, the $n \times n$ chessboard can be covered on a Klein bottle with $(\frac{1}{3}(n+2))^2 - k$ kings; that is, with k fewer kings than you need on a regular chessboard.

Now, this same idea doesn't work on an 8×8 chessboard because the resulting 3×16 band across the top and bottom needs six kings and you haven't saved anything. Unfortunately, this is true more generally and, on a Klein bottle, an $n \times n$ board where $n = 3k - 1$ needs just as many kings as a $3k \times 3k$ board needs, namely, k^2. It is therefore only when n is of the form $n = 3k + 1$ that you can do better on the Klein bottle than on the regular chessboard. It will prove useful, however, to further split this case $n = 3k + 1$ into two sub-cases: $n = 6k + 1$ and $n = 6k + 4$. The case $n = 6k + 1$ was dealt with in Problem 9.5.

When n is of the form $n = 6k + 4$, we have k bands of length $12k + 8$, taking three rows at a time from the top and three rows at a time from the bottom, and then the remaining four rows in the center also have to be treated as another band, two rows wide this time, of length $12k + 8$.

Each of these $k+1$ bands needs $4k+3$ kings, making a total of $(k+1)(4k+3) = 4k^2 + 7k + 3$ kings, which can be written as $(2k+2)^2 - (k+1) = (\frac{1}{3}(n+2))^2 - (k+1)$. Thus, you need $k+1$ fewer kings in this case than on the regular chessboard.

We can summarize things for square boards on the Klein bottle as follows:

$$\gamma(K_{n\times n}^{\text{klein}}) = \begin{cases} \frac{1}{9}n^2 & \text{for } n = 3k; \\ (\frac{1}{3}(n+1))^2 & \text{for } n = 3k+2; \\ (\frac{1}{3}(n+2))^2 - \frac{1}{6}(n-1) & \text{for } n = 6k+1; \\ (\frac{1}{3}(n+2))^2 - \frac{1}{6}(n+2) & \text{for } n = 6k+4. \end{cases}$$

I have given you the impression up to this point that one can fare somewhat better in terms of kings domination on a torus than one can on the Klein bottle. This is certainly the case for the square chessboards we have just considered, but if we turn now to rectangular chessboards, we will see that things are far more balanced in this regard than you may have originally thought.

The first thing to notice about a rectangular board on a Klein bottle is that it makes a difference which way we turn the chessboard. The following problem should make that point quite clear, and also give you a chance to see that the approach we used for square boards on a Klein bottle works just as easily for rectangular boards. For consistency, I will continue to maintain a convention that the twist in the Klein bottle will always occur between the left-hand edge and the right-hand edge of the chessboard.

Problem 9.6 Show that on a Klein bottle a 14×7 board can be covered with thirteen kings; whereas, a 7×14 board needs fifteen kings on a Klein bottle.

We can give a general result on kings domination for rectangular boards on the Klein bottle using this exact same approach and the result is given below. The details are to be found in [34]. You might have noticed that since no use has been made of the identification between the top and bottom edges of these chessboards on Klein bottles, we could equally well be thinking of these chessboards as being on Möbius strips.

Theorem 9.2 (Watkins and McVeigh) *The kings domination number on a Klein bottle for a rectangular $m \times n$ chessboard is given by*

$$\gamma(K^{\text{klein}}_{m\times n}) = \begin{cases} \left\lceil \frac{m}{6} \right\rceil \left\lceil \frac{2n}{3} \right\rceil - \left\lceil \frac{n-1}{3} \right\rceil, & \text{for } m \equiv 1, 2, 3 \bmod 6; \\ \left\lceil \frac{m}{6} \right\rceil \left\lceil \frac{2n}{3} \right\rceil, & \text{for } m \equiv 4, 5, 6 \bmod 6. \end{cases}$$

In particular, when $m = n$, these formulas reduce to the ones given above for $\gamma(K^{\text{klein}}_{n\times n})$.

Problem 9.7 Show that for the 6×7 and the 11×7 chessboards, you can do better on a Klein bottle covering these boards with kings than you can on a torus.

THE BISHOPS GRAPH ON A KLEIN BOTTLE

A remarkable thing happens to bishops on a Klein bottle, and this was first noticed by a student, Jennifer Johnson. In her words: 'what is most fascinating is what happens to the diagonals; it is as if each diagonal in each direction was doubled in length'. Consider the bishop in the left-hand diagram in Figure 9.6. If this bishop moves up and to the right along a positive diagonal, it will go off the board at a, reappear at the bottom and continue on, up and to the right, to b; there it goes off the board once again, but this time, because of the twist in the Klein bottle, it reappears on the left at b and is now moving down and to the right; when it reaches the bottom at a, it reappears at the top and continues on to c; there it once again reverses direction as it reappears on the left, and continues on up and to the right until it returns to the original square. You can see that it has, in fact, traced out two *full* diagonals—what I earlier called *toroidal* diagonals—and a total of sixteen squares. The same thing happens if the bishop moves along its negative diagonal, as shown in the middle diagram in Figure 9.6. The total scope of a single bishop on a Klein bottle, therefore, is 28 squares, as can be seen in the final, combined diagram on the right in Figure 9.6.

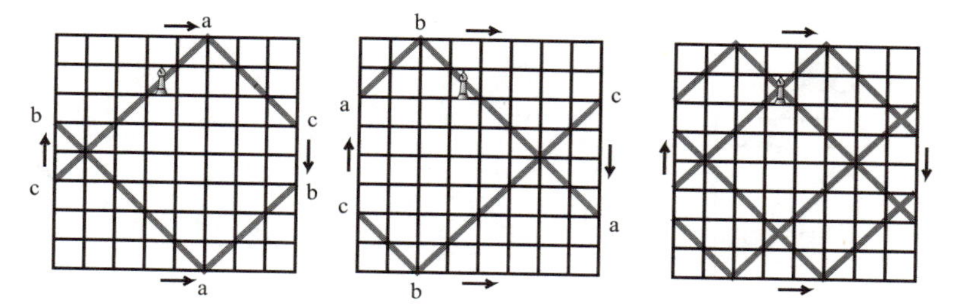

Figure 9.6 The two bishop's diagonals on a Klein bottle.

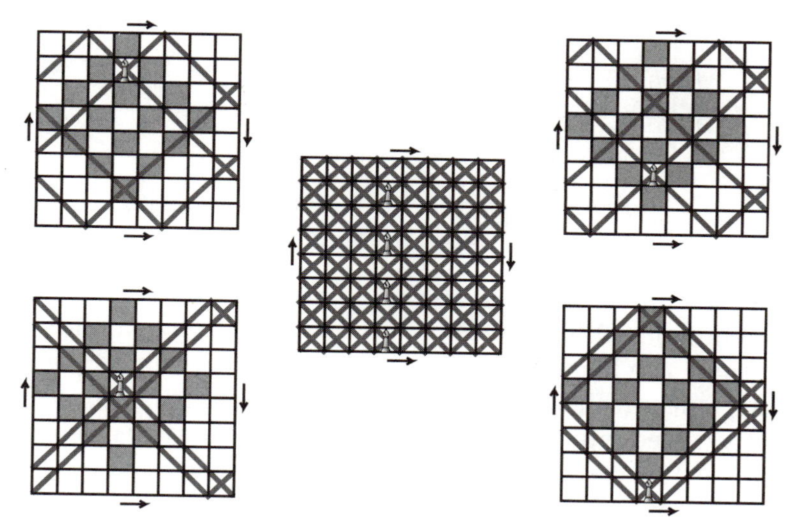

Figure 9.7 Covering the 8×8 chessboard on a Klein bottle with 4 bishops.

Since a bishop does cover twice as many squares on a Klein bottle as it covers on a torus, it is not at all surprising that we can prove the following result.

Theorem 9.3 (Johnson) *The bishops domination number for an $n \times n$ chessboard on a Klein bottle is given by*

$$\gamma(B_{n\times n}^{\text{klein}}) = \left\lceil \tfrac{1}{2}n \right\rceil.$$

In order to understand how this works in general it is useful to look carefully at the 8×8 chessboard. In Figure 9.7, we see in separate diagrams the squares that are covered by four bishops,

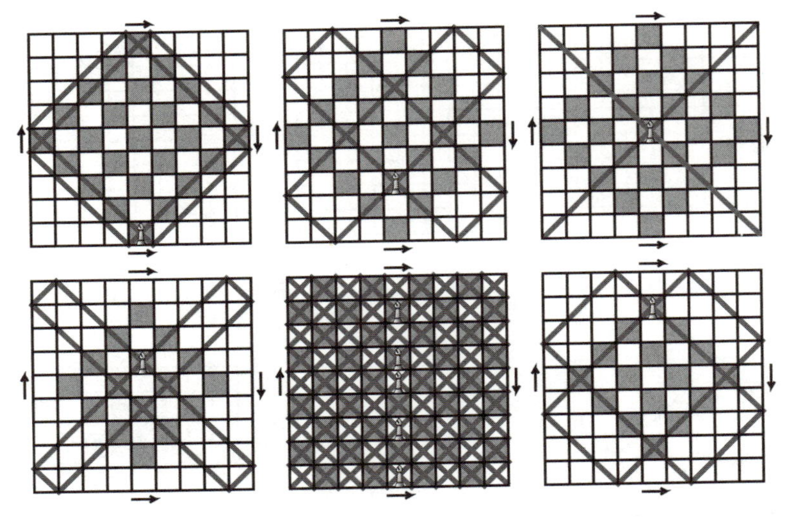

Figure 9.8 Covering the 9×9 chessboard on a Klein bottle with 5 bishops.

first individually, and then how they combine to cover the entire 8×8 chessboard. Thus, $\gamma(B_{8\times 8}^{\text{klein}}) \leqslant 4$. However, you should also note that each bishop crosses the shaded 4×4 square pattern exactly once in the positive direction, and exactly once in the negative direction—and this would be true no matter where you place a bishop on the board. Therefore, it is going to take at least four bishops just to cover these sixteen shaded squares, and hence, at least four bishops to cover the entire board. Therefore, $\gamma(B_{8\times 8}^{\text{klein}}) \geqslant 4$, and so $\gamma(B_{8\times 8}^{\text{klein}}) = 4$. This same idea works in general for any even value of n. And so, for n even, $\gamma(B_{n\times n}^{\text{klein}}) = \frac{1}{2}n$.

The case where n is odd can be illustrated by looking at the 9×9 chessboard. In Figure 9.8, we see in separate diagrams the squares that are covered by five bishops individually, and then how they combine to cover the entire 9×9 chessboard. Thus, $\gamma(B_{9\times 9}^{\text{klein}}) \leqslant 5$. However, you should also note that each bishop either crosses the shaded 5×5 square pattern at most twice in each direction—as in the top three diagrams in Figure 9.8—or crosses the shaded 4×4 square pattern twice in each direction—as in the two lower diagrams in Figure 9.8. Moreover, this will be true no matter where you place a bishop on the board. Therefore, it takes at least five bishops to cover just these shaded squares, and, hence, at least five bishops to cover the board. Note that three of these bishops are on white squares, and two of these bishops are on black squares. Therefore, $\gamma(B_{9\times 9}^{\text{klein}}) \geqslant 5$, and so $\gamma(B_{9\times 9}^{\text{klein}}) = 5$. This idea works in general for any odd value of n. And so, for n odd, $\gamma(B_{n\times n}^{\text{klein}}) = \frac{1}{2}(n+1)$.

CHAPTER 9

Solutions to Problems

Solution 9.1 Since a knight covers ten squares on the $2 \times 2 \times 2$ cube in addition to the square it occupies, two knights are not sufficient to cover the 24 squares on the cube. The arrangement of three knights shown in Figure 9.9, however, does manage to cover the entire cube.

Figure 9.9 A knight's cover of the $2 \times 2 \times 2$ cube.

At least four bishops will be required to cover the $2 \times 2 \times 2$ cube since a bishop covers only six squares in addition to the square it occupies. On the other hand, a single bishop covers a square on each of the six 2×2 faces of the cube, so there is at least a chance that four bishops will be enough. And it is. In fact, we can even put all four of them on one face, as is shown in Figure 9.10.

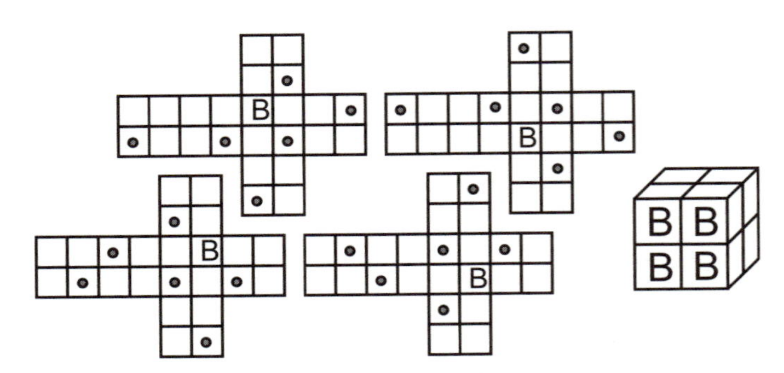

Figure 9.10 A bishop's cover of the $2 \times 2 \times 2$ cube.

A rook covers fourteen squares in addition to the square it occupies, and the arrangement of two rooks in Figure 9.11 easily covers the $2 \times 2 \times 2$ cube.

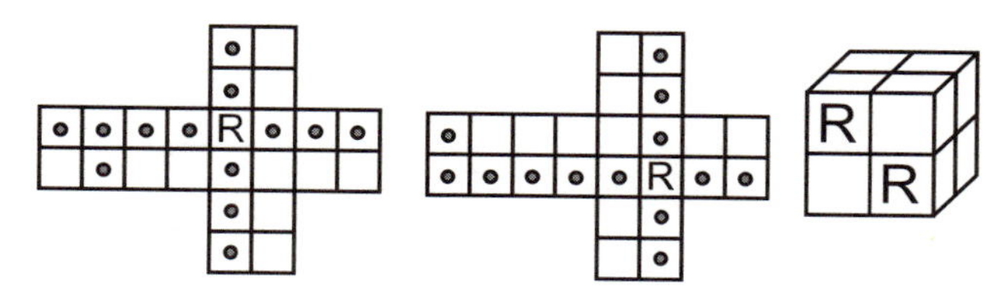

Figure 9.11 A rook's cover of the $2 \times 2 \times 2$ cube.

A queen covers seventeen squares in addition to the square it occupies, so you also need two queens to cover the $2 \times 2 \times 2$ cube, as shown in Figure 9.12.

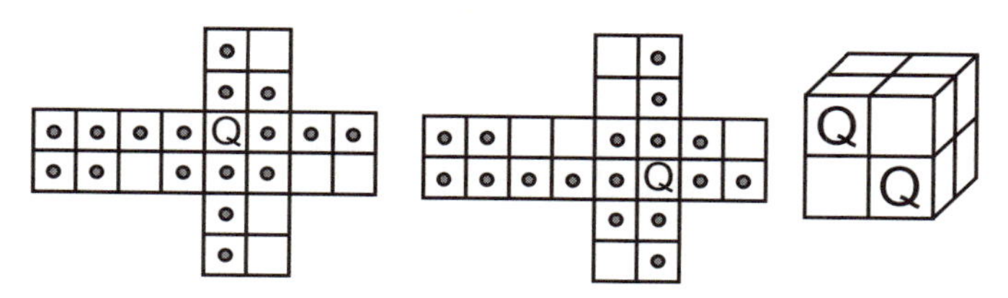

Figure 9.12 A queen's cover of the $2 \times 2 \times 2$ cube.

A king covers seven squares in addition to the square it occupies, that is, a total of eight squares. So it is conceivable that the entire $2 \times 2 \times 2$ cube could be covered with just three kings. However, it is clear that the relative geometry of these eight squares—being four squares on a single face together with the two immediately adjacent squares from each of two mutually adjacent faces—makes this impossible. Placing four kings with two kings on each of two opposing faces, as shown in Figure 9.13, does work, however.

Solution 9.2 The twelve squares are shown in Figure 9.14.

Solution 9.3 Figure 9.3, of course, shows the case $k = 0$ and $n = 7$. In general, this can be done by placing $k + 1$ kings in each row, as shown in Figure 9.15 for the $k = 1$ and $k = 2$ cases. Note that, in all three of these cases, from the position of the very

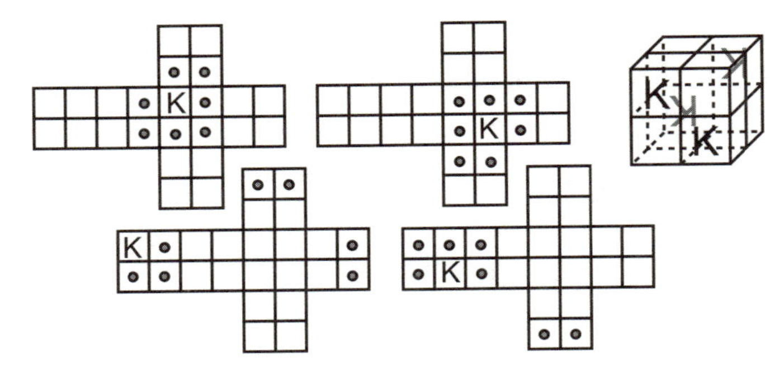

Figure 9.13 A king's cover of the $2 \times 2 \times 2$ cube.

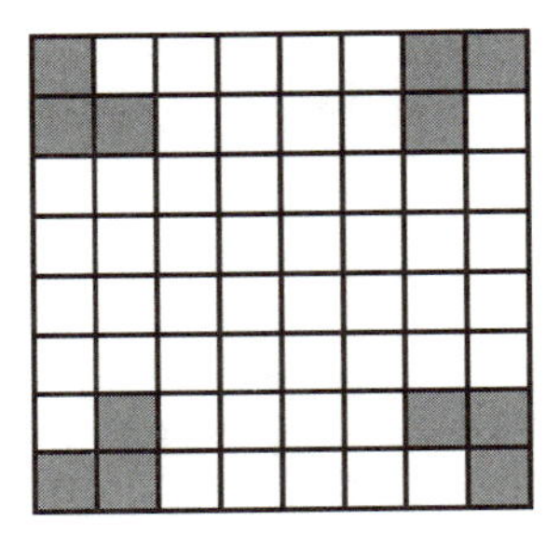

Figure 9.14 Twelve squares on a regular 8×8 chessboard that must be covered by knights.

last king at the bottom it is exactly three squares over and one square down to the first king back at the top. This is a critical detail and ensures that all of the squares near the end of the top row get covered, and, of course, this also happens in the general case because the overall pattern represents a shift of three squares to the right done exactly $n(k+1)$ times, and $3n(k+1) \equiv 0 \bmod n$.

Solution 9.4 For the 7×10 board,

$$\left\lceil \frac{7\lceil \frac{10}{3} \rceil}{3} \right\rceil = \left\lceil \frac{7(4)}{3} \right\rceil = 10 \quad \text{and} \quad \left\lceil \frac{10\lceil \frac{7}{3} \rceil}{3} \right\rceil = \left\lceil \frac{10(3)}{3} \right\rceil = 10;$$

and so 10 kings are required. The same procedure we used for square chessboards can be used, and the resulting arrangement is shown on the left in Figure 9.16.

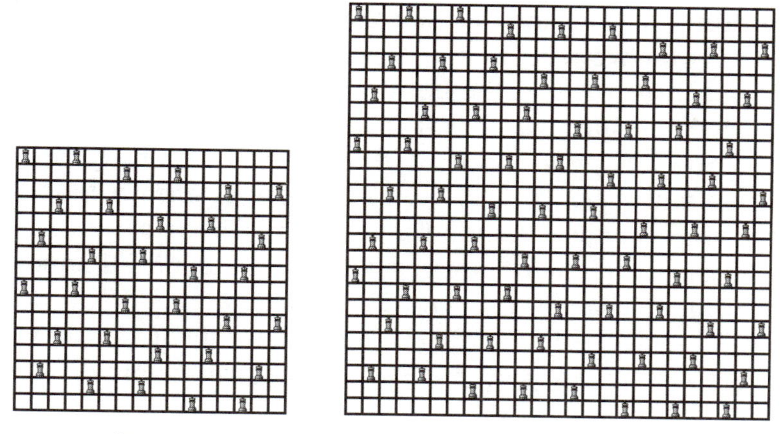

Figure 9.15 Kings dominating the 16×16 and the 25×25 toroidal chessboards.

For the 8×10 board,

$$\left\lceil \frac{8\lceil \frac{10}{3} \rceil}{3} \right\rceil = \left\lceil \frac{8(4)}{3} \right\rceil = 11 \quad \text{and} \quad \left\lceil \frac{10\lceil \frac{8}{3} \rceil}{3} \right\rceil = \left\lceil \frac{10(3)}{3} \right\rceil = 10;$$

and so 11 kings are required. However, the 8×10 board is also a special case—the case where $m = 3j - 1$, $n = 3k - 2$, $j \leqslant k$, and $j \geqslant \frac{1}{2}k$—and in this case the basic procedure needs a slight modification. If the board is put in a horizontal position, as shown in the middle of Figure 9.16, then the last king would need to be placed one square short of its normal position; otherwise, a square in the top row would remain uncovered. More simply, however, we can just rotate the board, and place the eleven kings as shown on the right in Figure 9.16.

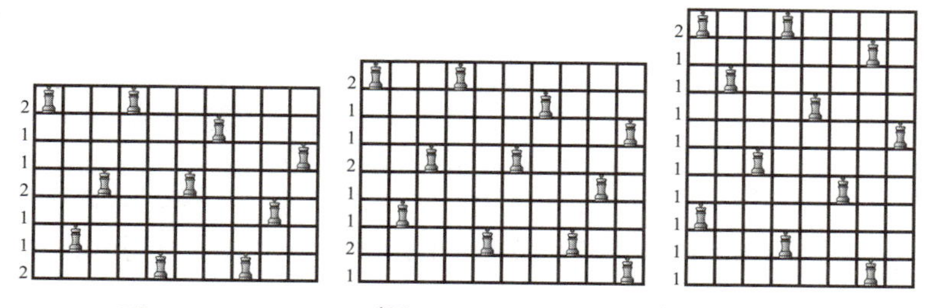

Figure 9.16 $\gamma(K^{\text{tor}}_{7\times 10}) = 10$ and $\gamma(K^{\text{tor}}_{8\times 10}) = 11$.

Solution 9.5 The idea is exactly the same as for the 7×7 board. You cover six rows at a time, taking three rows at the top and three rows at the bottom—working your way toward the center—which gives you a band of length $2(6k+1)$ to cover at each stage. Each of these bands can therefore be covered with $\lceil \frac{1}{3}(2(6k+1)) \rceil = 4k+1$ kings. Since $n = 6k+1$, there are k of these bands for us to do. This leaves a single row in the middle to be covered, which contain $6k+1$ squares, and so it can be covered with $2k+1$ kings. This is a total of $k(4k+1) + (2k+1) = 4k^2 + 3k + 1$ kings, which can be written as $(2k+1)^2 - k = (\frac{1}{3}(n+2))^2 - k$, as claimed. We illustrate this in Figure 9.17 for a 13×13 chessboard.

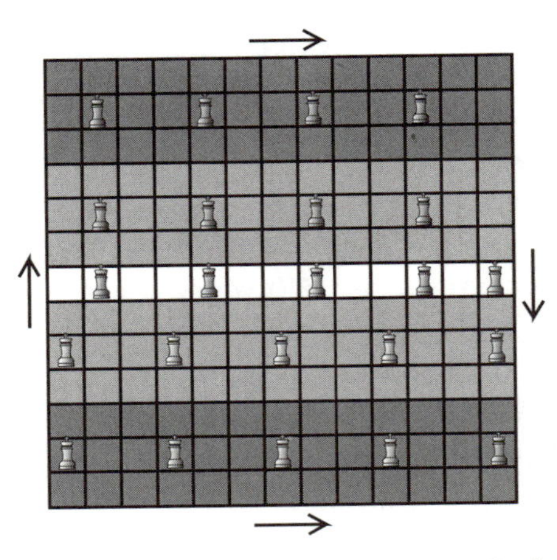

Figure 9.17 Kings dominating a 13×13 chessboard on a Klein bottle.

Solution 9.6 The difference between these two boards is that the 14×7 board has two bands of length fourteen that can be covered with five kings each, leaving only two rows in the middle to be covered with three additional kings, a total of thirteen kings; whereas, the 7×14 board has one band of length 28 that requires ten kings, and leaving a row in the middle which needs another five kings, a total of fifteen kings. This is shown in Figure 9.18.

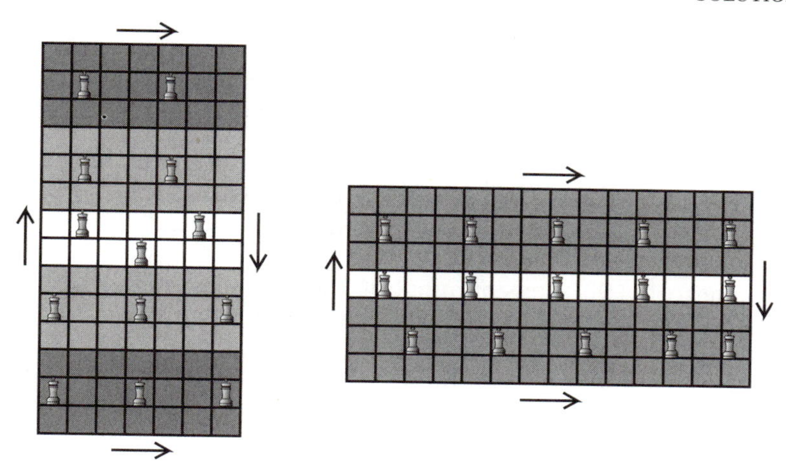

Figure 9.18 Kings dominating a 14×7 board and a 7×14 board on a Klein bottle.

Solution 9.7 For the 6×7 board on the torus, you have to compute

$$\left\lceil \frac{6\lceil \frac{7}{3} \rceil}{3} \right\rceil = \left\lceil \frac{6(3)}{3} \right\rceil = 6 \quad \text{and} \quad \left\lceil \frac{7\lceil \frac{6}{3} \rceil}{3} \right\rceil = \left\lceil \frac{7(2)}{3} \right\rceil = 5;$$

and take the maximum, which is 6. This says you need the same number of kings on a torus as you do for a regular 6×7 chessboard. For the Klein bottle, you compute

$$\left\lceil \frac{6}{6} \right\rceil \left\lceil \frac{2(7)}{3} \right\rceil = \left\lceil \frac{14}{3} \right\rceil = 5,$$

and so you can do better on a Klein bottle. The respective coverings are shown in Figure 9.19. This same thing will happen on boards of size 6×10, 6×13, 6×16, and so on, since for any $n = 3k + 1$, the $6 \times n$ board on a torus will need $2(k + 1)$ kings, and the same board on a Klein bottle will need only $2k + 1$ kings.

For the 11×7 board on the torus, you have to compute

$$\left\lceil \frac{11\lceil \frac{7}{3} \rceil}{3} \right\rceil = \left\lceil \frac{11(3)}{3} \right\rceil = 11 \quad \text{and} \quad \left\lceil \frac{7\lceil \frac{11}{3} \rceil}{3} \right\rceil = \left\lceil \frac{7(4)}{3} \right\rceil = 10;$$

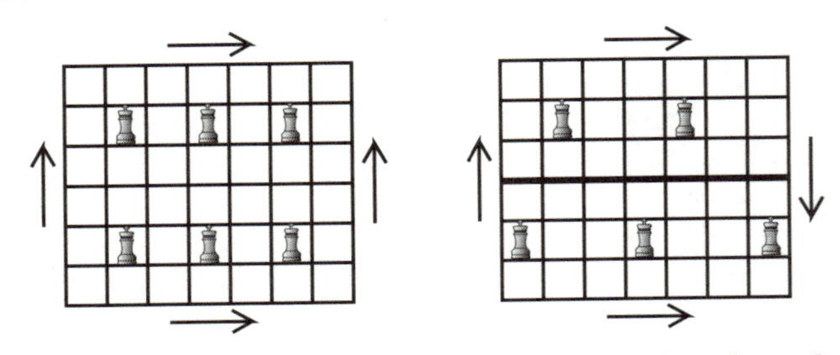

Figure 9.19 Kings dominating the 6×7 chessboard on a torus, and on a Klein bottle.

and take the maximum, which is 11. For the Klein bottle, you compute

$$\left\lceil \frac{11}{6} \right\rceil \left\lceil \frac{2(7)}{3} \right\rceil = 2(5) = 10,$$

and so, again, you can do better on a Klein bottle. The respective coverings are shown in Figure 9.20.

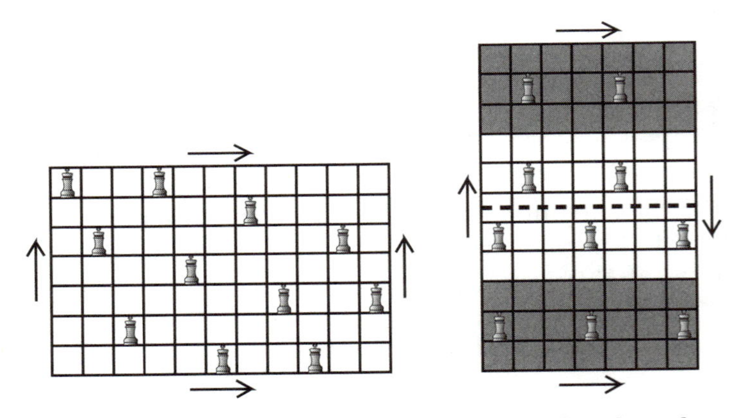

Figure 9.20 Kings dominating the 11×7 chessboard on a torus, and on a Klein bottle.

Independence

The concept of independence is closely related to that of domination, and is, in its own right, one of the central ideas in graph theory. We call two vertices in a graph *independent* if they are not adjacent, that is, if there is no edge joining them. This quite naturally gives rise to the following definitions.

Independent Sets

Definition 10.1 A set S of vertices in a graph G is said to be an *independent set* if no two vertices in S are adjacent. The *independence number* of a graph G is, then, the maximum size of an independent set in the graph G. We denote this independence number of the graph G by $\beta(G)$. An independent set with this maximum size is called a *maximum independent set.*

Just as we did in Chapter 7 for domination, we can take as an example the bishops graph $B_{4\times 4}$ for the 4×4 chessboard shown in Figure 10.1, where again the graph has two components, one for the white squares and one for the black squares. It is easy to see that in each component the maximum size of an independent set is 3. This is because, in each component, the four vertices in the center are all mutually adjacent, so only one of them can be included in an independent set; and, of the four remaining vertices on the outside, at most two can be included in an independent set. One choice for a maximum independent set for this graph would be the set $\{1, 5, 9, 2, 4, 14\}$. Therefore, $\beta(B_{4\times 4}) = 6$.

One thing to notice in this example is that it is quite obvious from the graph in Figure 10.1 that none of the vertices 7,

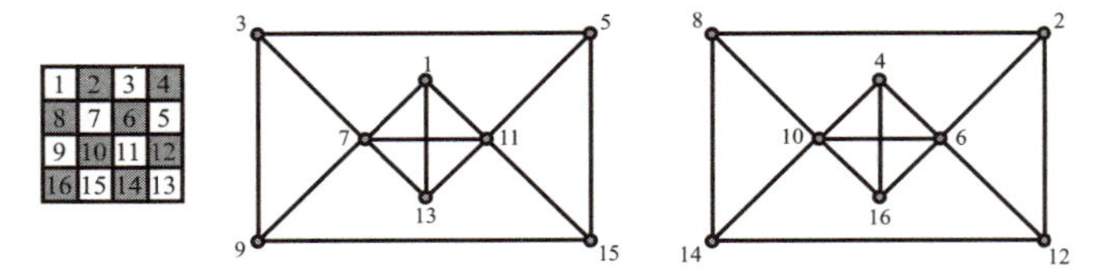

Figure 10.1 The bishops graph $B_{4\times4}$.

11, 10, or 6 can occur in a maximum independent set. In other words, in the language of chess problems, if we have a maximum number of independent bishops placed on a 4×4 chessboard, they must all be along the outer ring of edge squares. As we will see later, this remains true for larger chessboards as well.

Another thing to notice about our example is that this maximum independent set is also a dominating set. A moment's reflection will tell you that this is always true for any graph. If there were a vertex in a graph that was not adjacent to any vertex in a maximum independent set—that is, if a maximum independent set failed to dominate the graph—then that vertex could simply be added to the maximum independent set, thereby creating a larger independent set than the maximum independent set, which is nonsense. Thus, for any graph, we always have that

$$\gamma(G) \leqslant \beta(G).$$

Since we are about to spend a considerable amount of time thinking about independent chess pieces, it would be wise to pause for a moment to inquire exactly how likely it is that two chess pieces, two queens say, are actually independent.

Problem 10.1 What is the probability that two randomly placed queens on an 8×8 chessboard are independent?

The 8-Queens Problem

The oldest, and most famous, independence problem is the *8-queens problem.* Its first appearance was in a German chess newspaper, the *Schachzeitung*, in September 1948, in a problem posed under a pseudonym by chess player Max Bezzel. The problem asked how to place eight queens on an 8×8 chessboard so that none of them attacks any of the others. By 1854, a total of 40 different solutions to this problem had been published in the *Schachzeitung*.

The problem was also posed in another German newspaper by Franz Nauck in June 1850, and Nauck correctly published the 92 possible solutions in the same newspaper on September 21, though without any proof that his list was complete. An interesting aside is that, in the summer of 1850, Carl Friedrich Gauss read Nauck's account of this problem and, by September 1, wrote to a friend that he had found 76 solutions. More correspondence ensued over the next month as the number of solutions was adjusted up and down, but Gauss did pass along to his friend, although apparently without much conviction, Nauck's complete list of 92 solutions. In typical Gaussian fashion, however, none of this came to light until much later, in 1865.

I urge you to read Paul Campbell's brief essay, *Gauss and the eight queens problem*, not only for its rather detailed history of the 8-queens problem, but also for the benefit of the increased awareness that his cautionary tale—which he subtitled: *A study in miniature of the propagation of historical error*—brings us concerning the way in which utterly unreliable information gets repeated over and over again once it makes its way into the historical mathematical record [5].

It is not surprising that Gauss turned the 8-queens problem into a problem about arithmetic in a rather interesting way. First of all, it is obvious that any solution to the 8-queens problem has one queen in each column, and one queen in each row; and so, a solution can be represented as a permutation of the numbers 1 through 8. For example, the permutation 1 5 8 6 3 7 2 4 can be used to represent the eight-queens solution shown in Figure 10.2.

Gauss devised an easy way to check a permutation to see if it represented an arrangement of independent queens, that is, to see if any two queens occupy the same diagonal. He illustrated his method with the permutation 1 5 8 6 3 7 2 4 from Figure 10.2

Figure 10.2 A solution to the 8-queens problem.

by performing the following sums:

$$\begin{array}{cccccccc} 1 & 5 & 8 & 6 & 3 & 7 & 2 & 4 \\ 1 & 2 & 3 & 4 & 5 & 6 & 7 & 8 \\ \hline 2 & 7 & 11 & 10 & 8 & 13 & 9 & 12 \end{array}$$

$$\text{and} \quad \begin{array}{cccccccc} 1 & 5 & 8 & 6 & 3 & 7 & 2 & 4 \\ 8 & 7 & 6 & 5 & 4 & 3 & 2 & 1 \\ \hline 9 & 12 & 14 & 11 & 7 & 10 & 4 & 5 \end{array}.$$

Note, in each case, that the eight sums are distinct integers. For the eight sums above, this tells us that no two queens lie on the same negative diagonal, while for the eight sums below, this tells us that no two queens lie on the same positive diagonal. Therefore, the queens represented by the permutation 1 5 8 6 3 7 2 4 are independent.

In *Mathematical Recreations*, Maurice Kraitchik gives all 92 solutions to the 8-queens problem [20]. They can be listed as follows: there are eleven permutations,

1 5 8 6 3 7 2 4, 1 6 8 3 7 4 2 5, 2 4 6 8 3 1 7 5, 2 5 7 1 3 8 6 4,
2 5 7 4 1 8 6 3, 2 6 1 7 4 8 3 5, 2 6 8 3 1 4 7 5,
2 7 3 6 8 5 1 4, 2 7 5 8 1 4 6 3, 3 5 8 4 1 7 2 6, 3 6 2 5 8 1 7 4,

each of which gives rise to eight different arrangements when all rotations and reflections are included; and one additional permutation,

3 5 2 8 1 7 4 6,

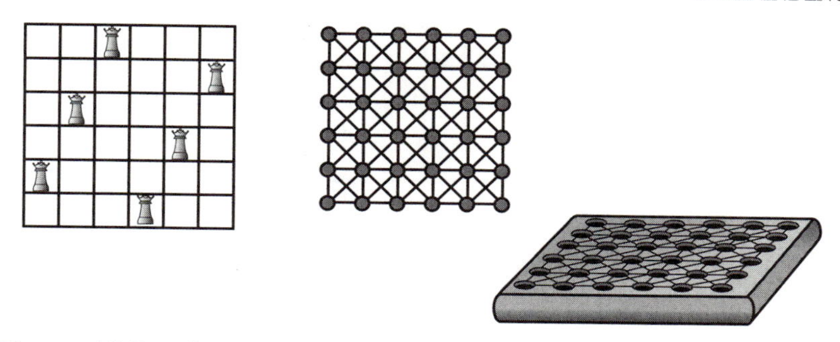

Figure 10.3 The 6-queens solution and a related London puzzle.

which is symmetric, and which gives rise to four different arrangements.

Problem 10.2 Use the permutation method of Gauss to find all solutions to the 5-queens problem. That is, find all possible arrangements of five independent queens on a 5×5 chessboard. Also, group your solutions according to their symmetry type.

Now, for the 6-queens problem, there are only four solutions, all based on a single symmetric solution 2 4 6 1 3 5, shown in Figure 10.3. As reported by W. W. Rouse Ball, this essentially unique solution was the basis for a wooden puzzle—illustrated in Figure 10.3—that was sold on the streets of London for a penny. The object of the puzzle was to place six wooden pegs into holes so that no two pegs would be in a straight line.

Note that, in the solution to the 8-queens problem in Figure 10.2, three of the queens—the queens at squares $(2, 5)$, $(4, 6)$, and $(6, 7)$—lie in a straight line, although, obviously, not in a line which corresponds to an actual diagonal, row, or column on the chessboard. H. E. Dudeney pointed out that, among the twelve fundamental solutions to the 8-queens problem, the only solution in which three queens do not lie along such a straight line is the one shown in Figure 10.4, namely, 4 2 8 5 7 1 3 6. Here is a related problem, which I'll present to you in two versions.

Problem 10.3 Can you place sixteen queens on an 8×8 chessboard so that no three queens lie in a straight line? When

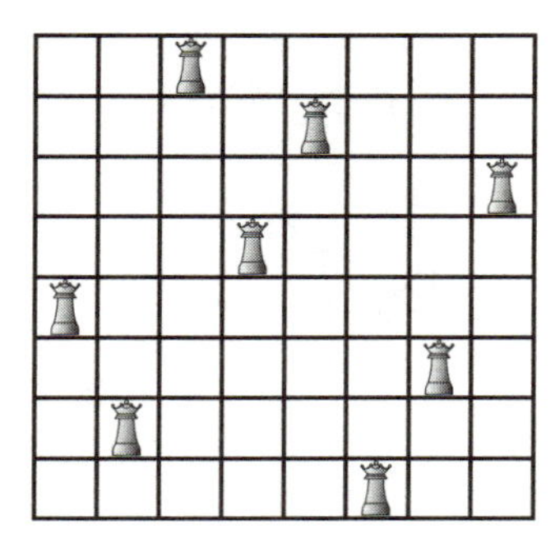

Figure 10.4 An 8-queens solution without three queens in a straight line.

you have succeeded in doing that, try Dudeney's original version [10], which he gave using pawns in order to avoid the obvious potential for confusion between the more numerous straight lines of geometry and the limited vertical, horizontal, and diagonal lines on a chessboard: can you place sixteen queens on an 8×8 chessboard so that no three queens lie in a straight line and the first two queens are placed at squares $(4, 4)$ and $(5, 5)$? The interesting wrinkle of restricting the placement of the first two queens to center squares was in fact introduced by Nauck in his original newspaper article on the 8-queens problem.

For a solution to the 7-queens problem, as well as to show you that even after 150 years the 8-queens problem is still causing people to come up with new and interesting ideas, I'd like to show you a remarkable new method for finding independent queens using magic squares [9]. This method is described by Clifford Pickover in *The Zen of Magic Squares, Circles, and Stars*, and uses the odd-order magic square construction of Muhammad ibn Muhammad and Bachet we saw in Chapter 4 in the form of a variation that is due to de la Loubère [25].

The first step in this method is to construct a 7×7 magic square using the de la Loubère variation, as shown at the upper-left in Figure 10.5. Note that this variation begins in the top row in the center, then moves *up* and to the right, and then drops down a single square when each diagonal is completed. The next step is to produce a new 7×7 array of numbers by

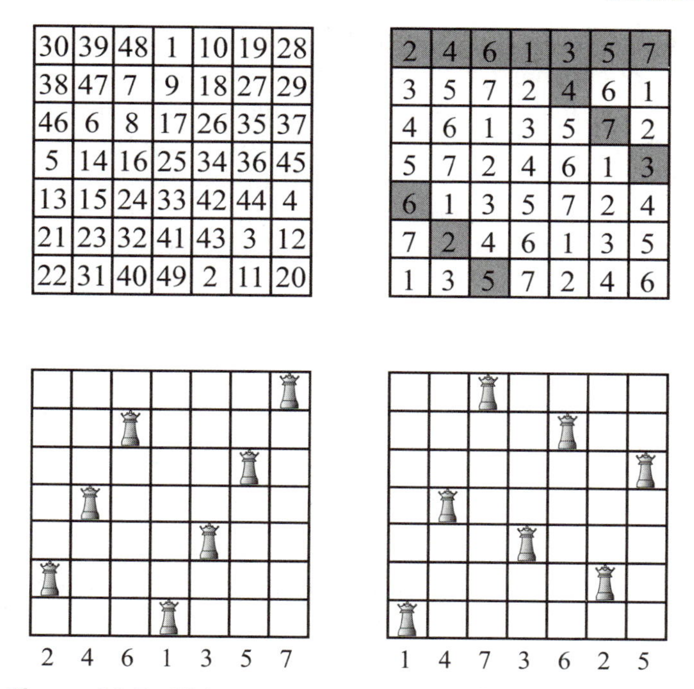

Figure 10.5 Using a magic square to find solutions for the 7-queens problem.

reducing, modulo 7, each number in the 7×7 magic square to get a new number from 1 to 7. This new array, shown at the upper-right in Figure 10.5, now contains solutions to the 7-queens problem in each row, and in each negative diagonal. For example, the top row—shaded in the figure—contains the solution 2 4 6 1 3 5 7; and the negative diagonal—also shaded in the figure—that begins at the top in the center contains the solution 1 4 7 3 6 2 5. Each of these 7-queens solutions is also shown in Figure 10.5.

THE n-QUEENS PROBLEM

It is inevitable that a problem as inherently interesting as the 8-queens problem would quickly evolve into an *n-queens problem*, which asks how to place n queens on an $n \times n$ chessboard so that none of them attacks any of the others. It is, of course, clear that you can't place more than n independent queens on an $n \times n$ chessboard, but it is not at all clear that you can always successfully place n independent queens on a board. So, our first order of business is to verify the following theorem of Wilhelm Ahrens, who wrote a wonderful book in 1901 on mathematical entertainment and games, *Mathematische Unterhaltungen und Spiele* [1] that even today contains a wealth of information on problems such as the 8-queens problem, knight's tours, magic squares, and on a wide range of other mathematical topics.

Theorem 10.1 (Ahrens) $\beta(Q_{2\times 2}) = 1$; $\beta(Q_{3\times 3}) = 2$; *otherwise, for* $n \neq 2, 3$,

$$\beta(Q_{n\times n}) = n.$$

Proof. I will follow the proof given by Yaglom and Yaglom [40] in 1964, which in turn is quite similar to Ahrens' original proof from 1910. The values for $n = 1$, 2, and 3 are obvious, so we begin the proof for $n \geqslant 4$. Since we will have to provide a construction for an independent set of queens for each value of n, it is not surprising that we will need to consider the odd values and the even values of n separately.

When n is even there is a very simple construction suggested by the solution to the 6-queens problem given in Figure 10.3. Start by placing a queen at $(1, 2)$, and do repeated knight's moves, one square over and two squares up, until you reach the top row. At this point you have occupied all of the columns in the left half of the board as well as all of the even-numbered rows. You now come back down to the bottom row and start again, one column over from where you left off, doing repeated knight's moves until you finish in the last column. This occupies the columns in the right half of the board as well as all of the odd-numbered rows. This simple construction is illustrated in Figure 10.6 for the 10×10 and 12×12 chessboards. Note that the arrangement for the right half of the board could have also been produced by merely turning the board around; in other words, the two halves are symmetric about the center of the board.

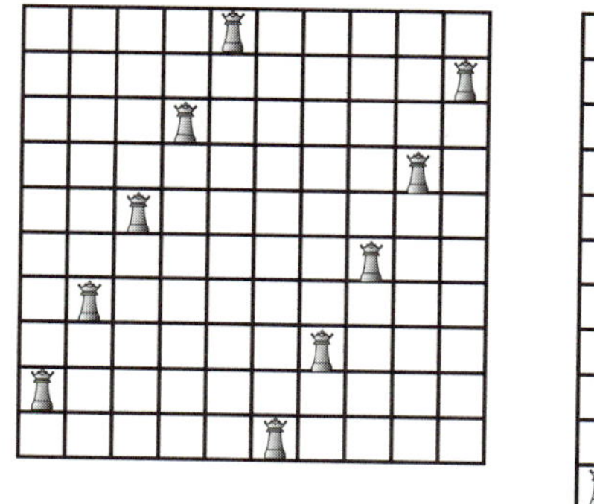
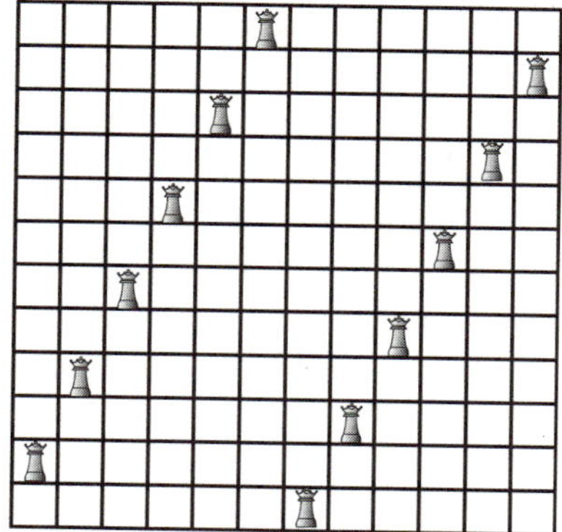

Figure 10.6 The basic construction for the n-queens problem when n is even.

Before continuing with the proof of the theorem, you may want to do the following problem.

Problem 10.4 Show that the construction pictured in Figure 10.6—a construction that works so nicely for the 6×6, 10×10, and 12×12 boards—fails for the 8×8 and 14×14 boards.

Based on what we have seen so far, we suspect that this construction works for all even n except when $n \equiv 2 \bmod 6$. This turns out to be the case, as we now show. It is clear that all we have to worry about in this construction are the negative diagonals. To be specific, since we can write $n = 2k$, we have to worry about whether an rth queen in the first group of k queens on the left could be on the same negative diagonal as an sth queen in the second group of k queens on the right.

The rth queen is at square $(r, 2r)$ and satisfies the equation $x + y = 3r$, whereas the sth queen occupies a square that has coordinates $(k + s, 2s - 1)$. So, if these two queens are on the same negative diagonal, then $(k + s) + (2s - 1) = 3r$, that is, $3s + k - 1 = 3r$. But, since $n = 2k$, this becomes $n = 6(r - s) + 2$.

Therefore, exactly as we suspected, this construction works except when $n \equiv 2 \bmod 6$, in which case it fails completely. So, for the case $n = 8, 14, 20, \ldots$, we need to do something else. Here is a construction that does the job in that case.

Start by skipping both the first column and the first row and placing the first queen at $(2, 3)$, and then doing repeated knight's moves, one square over and two squares up, as before, until you have to stop when you reach the row immediately below the top row. So far so good, but the first column and the first row are still unoccupied. To take care of that, we go back to the next-to-last knight that we just placed on the board—the one in the fourth row from the top—and *clone it* into two copies of itself, moving one copy all the way to the left and into the first column, and moving the other copy all the way down and into the first row.

At this point in the construction, we have placed k queens on the board, and they occupy all of the columns in the left half of the board, as well as all of the odd-numbered rows. For the right half of the board, for each queen already in place on the left, simply place a queen in the corresponding symmetric position on the right side with respect to the center of the board. In other words, rotate the board 180° and repeat. This construction is illustrated in Figure 10.7 for the 8×8 and the 14×14 chessboards, where the shaded squares show you the locations where the cloning took place. Note that, because this construction explicitly relies upon symmetry, it produced as a solution for the 8×8 board the only solution among the twelve fundamental solutions which has symmetry.

To verify that this construction indeed works for all n where $n \equiv 2 \bmod 6$, we begin the proof at the point in the construction when $k - 1$ of the queens have been placed on each side, but before any cloning has taken place. Imagine, then, if you will, that in Figure 10.7 the cloned queens in the left column and the bottom row of each board have coalesced back into a single queen at the shaded square from which they first materialized, and that, similarly, the cloned queens in the right column and the top row of each board have also coalesced back into a single queen at the shaded square from which they came.

At this imaginary point in the construction, then, there are $n - 2$ queens placed on the board. To verify that these $n - 2$ queens are independent, it is clear that we only need to check that no two of these queens, one from the first group on the

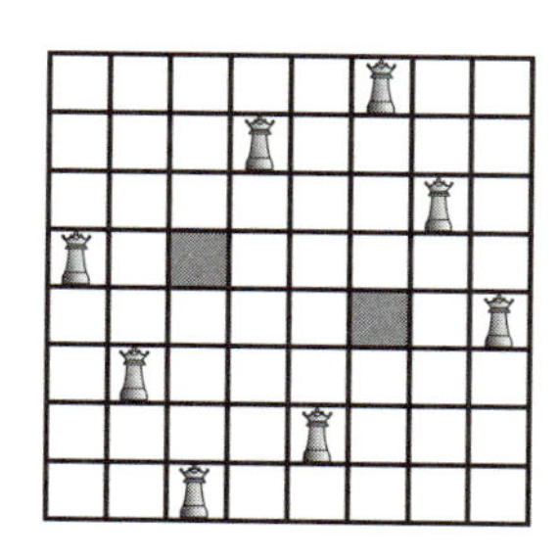

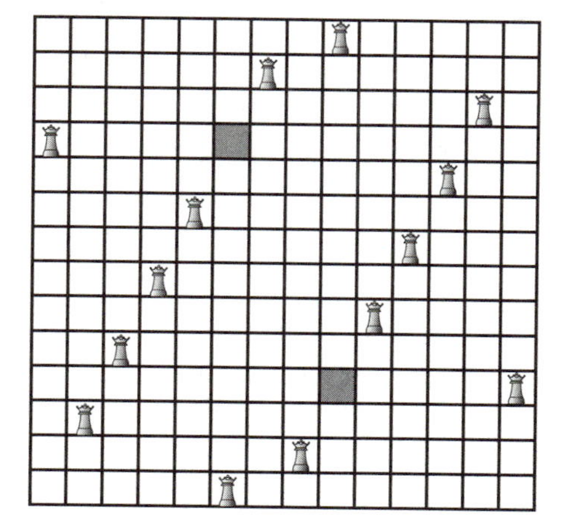

Figure 10.7 A special construction for the n-queens problem when $n \equiv 2 \bmod 6$.

left and one from the second group on the right, occupy the same negative diagonal. With this in mind, we see that an rth queen in the first group is at square $(r+1, 2r+1)$, whereas an sth queen in the second group is at square $(k+s, 2s)$. If these two queens are on the same negative diagonal, then they each must satisfy the same linear equation, $x + y = b$, for some integer b. Thus, $(r+1) + (2r+1) = (k+s) + (2s)$; that is, $3r + 2 = 3s + k$, which reduces to $n = 6(r-s) + 4$ since $n = 2k$. But, since we also know that $n \equiv 2 \bmod 6$, this would be a contradiction, so no two queens could have been on the same negative diagonal. Therefore, these $n-2$ queens are independent.

Moreover, it is clear that the cloning step itself does not cause any problem with independence as far as rows or columns are concerned. We do need to check the diagonals, however. The 'left' clone is at square $(1, n-3)$. If this queen is on the same negative diagonal as an rth queen in the first group, then $n - 2 = 3r + 2$, and so $n = 3r + 4$, a contradiction, since $n \equiv 2 \bmod 6$. If the cloned queen at $(1, n-3)$ is on the same negative diagonal as an sth queen in the second group, then $n - 2 = 3s + k$, which, since $n = 2k$, reduces to $k = 3s + 2$, another contradiction, since this would mean that $n = 2k = 6s + 4$. The 'bottom' clone is at square $(k-1, 1)$ and is easier

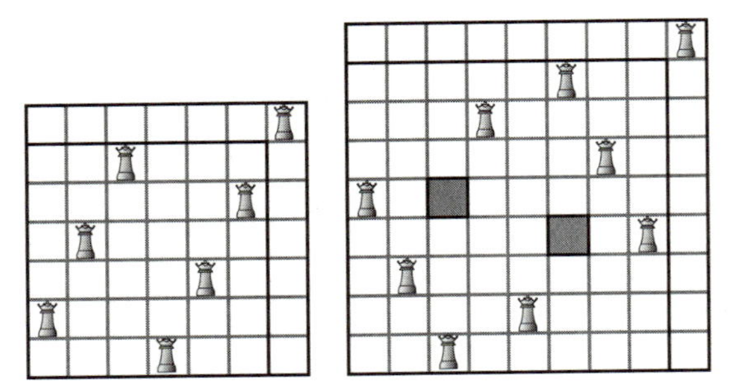

Figure 10.8 The construction for the n-queens problem when n is odd.

for us to handle, since it would be a problem only if $k = 3r + 2$, that is, if $n = 2k = 6r + 4$, which can't happen. It then follows by symmetry that the 'right' clone and the 'top' clone are each independent as well.

This completes the proof of the theorem for even boards. What about odd boards? The construction for them is now a piece of cake. Both of the constructions we used for even boards left the main positive diagonal empty. Therefore, in order to do an odd board, all we have to do is take a previously constructed even board, add a row at the top and a column on the right, and then place a single queen in the upper right-hand corner. This is illustrated in Figure 10.8 for the 7×7 board and the 9×9 board. This completes the proof of Theorem 10.1. □

In one sense, then, Theorem 10.1 has solved the n-queens problem in that $\beta(Q_{n\times n})$ has been determined for all values of n. In a wider sense, however, the n-queens problem remains far from answered. Historically, the 8-queens problem didn't really just ask whether there was a solution with eight queens, since this was rather easy to solve, but rather how many solutions there were and how to find them all. Maurice Kraitchik addressed the n-queens problem in this larger sense in a serious way for small values of n in his book, *Mathematical Recreations* [20]. There, he classified solutions into the following three different categories, according to their degree of symmetry.

Ordinary solutions (O): these are solutions that have no symmetry and therefore give rise to a total of eight solutions under rotation and reflection.

Centrosymmetric solutions (C): these are solutions that are unchanged under rotations of 180°, but are changed by other rotations or reflections, and therefore give rise to a total of four solutions.

Doubly centrosymmetric solutions (Q): these are solutions that are unchanged under any rotation, but are changed by reflections, and therefore give rise to two solutions.

An example of an ordinary solution is one like 1 5 8 6 3 7 2 4 in Figure 10.2; a centrosymmetric solution is one like 5 3 1 7 2 8 6 4 in Figure 10.7; and, an example of a doubly centrosymmetric solution for a 5×5 board is 2 5 3 1 4 in Figure 10.16.

Kraitchik even worked out the following data on the number of solutions for an $n \times n$ board, where for each n the total number of solution is given by:

$$\text{total} = 8Q_n + 4C_n + 2Q_n.$$

n	O_n	C_n	Q_n	total
4			1	2
5	1		1	10
6		1		4
7	4	2		40
8	11	1		92
9	42	4		342
10	89	3		724
11	329	12		2,680
12	1,765	18	1	14,200
13	?	31	1	
14	?	103		
15	?	298		

It is perhaps worth pausing to mention that these classifications are best described and best understood using the language of *group theory*, a subject which was first glimpsed by Evariste Galois around 1830 in his study of the theory of equations, and then later, in the 1870s, applied by Felix Klein to the study of geometry in the following very natural way.

The geometry of a square can be captured and 'algebraically' described, in terms of what is called its *symmetry group*, because a square has natural symmetries that we can easily classify. It turns out that for a square we can use just two basic symmetries to describe all of the rest: a clockwise rotation, ρ, of 90°, and a reflection, σ, about the central vertical axis of the square. So, for example, a clockwise rotation of 180° then becomes ρ^2, whereas a *counter*clockwise rotation of 90° would be represented by ρ^{-1}, which is obviously the same as ρ^3; or a reflection about the central horizontal axis of the square would be $\sigma\rho^2$, whereas, a reflection about the main positive diagonal of the square would be $\rho\sigma$. Note that, in this notation, we are reading right to left, and that $\rho\sigma$ and $\sigma\rho$ are different.

The *symmetry group of the square* then consists of

$$\{i, \rho, \rho^2, \rho^3, \sigma, \sigma\rho, \sigma\rho^2, \sigma\rho^3\},$$

where i is called the identity, which corresponds to no rotation or reflection at all. This group then has various subgroups, that is to say, subsets that are themselves groups. For example, $\{i, \rho, \rho^2, \rho^3\}$ forms a subgroup, since any combination of these rotations simply gives another rotation.

Kraitchik's doubly centrosymmetric solutions (Q) are, then, simply those solutions that are invariant under the subgroup $\{i, \rho, \rho^2, \rho^3\}$, but not under any other element of the group. The centrosymmetric solutions (C) are those solutions that are invariant under the subgroup $\{i, \rho^2\}$, but not under any other element of the group. Finally, the ordinary solutions (O) are those solutions that are invariant under the subgroup $\{i\}$, but not under any other elements of the group.

Pólya's Doubly Periodic Solutions

In 1918 George Pólya introduced what he called *doubly periodic* solutions to the n-queens problem, and he showed that such

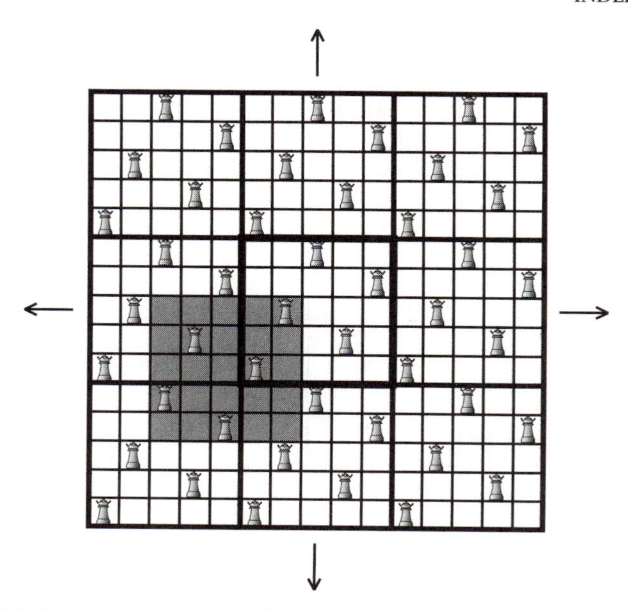

Figure 10.9 A doubly periodic solution to the 5-queens problem.

solutions exist if and only if $n \equiv 1$ or 5 mod 6. For example, consider the solution 1 3 5 2 4 for the 5×5 board. The queens are at squares (1,1), (2,3), (3,5), (4,2), (5,4). If we shift all the queens +2 in the y direction modulo 5—that is, as if we were on a torus—then the queens will now be at squares (1,3), (2,5), (3,2), (4,4), (5,1), which is another solution for the 5×5 board, as you can see in Figure 10.16. Alternatively, had we shifted all the queens +3 in the x direction modulo 5, then the queens would be at squares (4,1), (5,3), (1,5), (2,2), (3,4), and this is also one of the solutions shown in Figure 10.16. Now, of course, the point of the name 'doubly periodic' is that we can do both of these shifts at the same time and still get a solution, in this case: (4,3), (5,5), (1,2), (2,4), (3,1), also to be found in Figure 10.16.

Now, an obvious construction to use when $n \equiv 1$ or 5 mod 6 for producing a solution is to simply use a repeated knight's move beginning, say, at $(1, 1)$. Proving that this indeed yields a solution and that it is doubly periodic is basically the same as—although, in fact, a bit easier than—the argument we used to check the proof of Theorem 10.1. The nature of doubly peri-

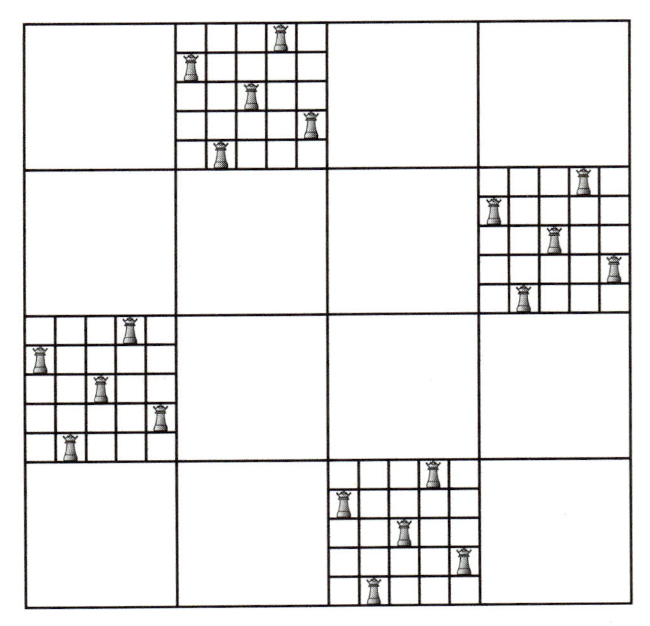

Figure 10.10 Using a doubly periodic 5×5 solution to solve the 20×20 board.

odic solutions can best be seen by replicating the $n \times n$ board and its solution in the plane as shown in Figure 10.9 for the 5×5 board and the solution 1 3 5 2 4. The original solution is shown in the center 5×5 square. Note that the solution corresponding to the double shift of +3 in the x direction and +2 in the y direction is to be found in the highlighted square, which itself has been shifted 3 to the left and 2 down from the original square. In other words, we see that the original solution was called *doubly periodic* because no matter where we now choose to place a highlighted 5×5 square upon this infinite plane it will be guaranteed to contain a solution to the 5-queens problem.

Pólya then went on to show that, in a very ingenious way, he could use a doubly periodic solution for an $n \times n$ chessboard together with any old solution for an $m \times m$ chessboard to produce a solution for an $mn \times mn$ chessboard. This idea is illustrated in Figure 10.10, where I have selected a doubly periodic solution for the 5×5 board and combined it with the decidedly *non*-doubly periodic solution for the 4×4 board from Figure 10.17. Note that it is precisely the doubly periodic nature—i.e. the *toroidal* nature—of the $n \times n$ solution that keeps all of the queens independent once they are replicated on the $mn \times mn$ board.

There is much work that is still being done on the n-queens problem, and in their excellent updated survey, *Combinatorial Problems on Chessboards*, Hedetniemi, Hedetniemi, and Reynolds give a brief, but highly detailed, description of current progress on the n-queens problem, work that, not surprisingly, frequently makes serious and interesting use of efficient techniques in computing [18].

Let us now turn to other chess pieces—the rook, the knight, the bishop, and the king—and see what is known about independence for each of these pieces. We begin with the easiest.

Independent Rooks

It is by now of course clear to us that any arrangement of independent rooks can be represented by a permutation, and vice versa, and so $\beta(R_{n \times n}) = n$, and the total number of arrangements is $n!$. This is illustrated in Figure 10.11 for the 3×3 chessboard and the $3! = 6$ permutations: 1 2 3, 1 3 2, 2 1 3, 2 3 1, 3 1 2, 3 2 1.

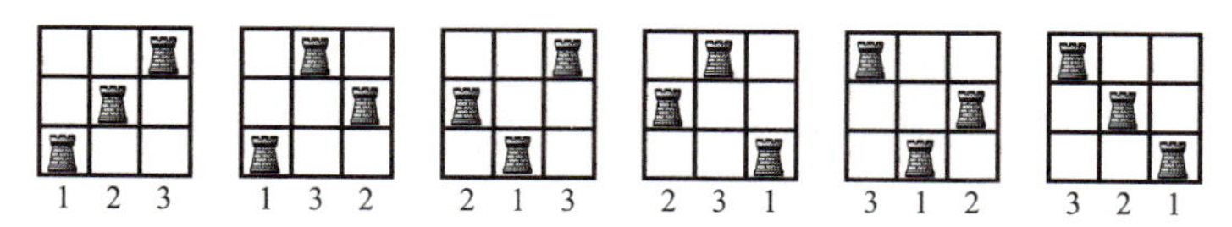

Figure 10.11 Independent rooks as the permutations: 1 2 3, 1 3 2, 2 1 3, 2 3 1, 3 1 2, 3 2 1.

This correspondence between independent rooks and permutations allows us to state an old problem in a new form, as it appears in [24].

Problem 10.5 Number the squares of an 8×8 chessboard from 1 to 64 in the normal fashion beginning with 1 in the upper left-hand corner. Show that no matter where you place eight independent rooks on the chessboard, the sum of the occupied squares will always be the same.

INDEPENDENT KNIGHTS

The independence problem for knights was first discussed in Chapter 1 where we observed that $\beta(N_{8\times 8}) = 32$. Since the knights graph is bipartite, 32 independent knights can be placed on the black squares or, equivalently, they can be placed on the white squares. This is the best you can do. Yaglom and Yaglom proved this by dividing the chessboard into eight 2×4 rectangles each of which can obviously contain at most 4 independent knights, which shows that $\beta(N_{8\times 8}) \leqslant 4(8) = 32$. An alternative proof of this was given by Ralph Greenberg in *The American Mathematical Monthly* in February 1964 using the fact that the 8×8 chessboard has a knight's tour—a tour which necessarily alternates between black and white squares—and, since independent knights cannot exist at adjacent squares anywhere along this tour, there can be at most 32 independent knights along the tour. Greenberg's proof has a significant bonus. The only way to pack 32 knights along a tour, then, is to alternate them as you go along the tour—that is, they must either *all* go on the white squares, or *all* go on the black squares.

It is pretty clear, then, what $\beta(N_{n\times n})$ is going to be in general: continue to put all the knights on either the white squares or the black squares when n is even, and on whichever of the two colors has the most squares in the case when n is odd—normally we arrange this to be the white squares. There does remain the awkward detail of showing that this strategy is the best that you can do, a detail that has often been overlooked. As Martin Gardner points out, we know that this is the best we can do as long as we know that a knight's tour exists or, in the odd case, as long as an open tour exists [14]. Fortunately, we have Schwenk's Theorem, Theorem 3.1, that tells us exactly when we have knight's tours, at least for the even case. We will then be able to finesse the odd case by combining both of the ideas that were used above for the 8×8 chessboard. Here is the result we are after.

Theorem 10.2 *$\beta(N_{2\times 2}) = 4$; otherwise, $\beta(N_{n\times n}) = \frac{1}{2}n^2$ for n even, and $\beta(N_{n\times n}) = \frac{1}{2}(n^2 + 1)$ for n odd.*

Proof. We'll do the even case first. For the 2×2 board, place a knight on each square. For the 4×4 board, not only is there not a knight's tour to use, but, by Problem 3.4, there is not even an open knight's tour. However, the 4×4 board easily divides into two 2×4 rectangles, each of which can contain at most 4 independent knights, and so $\beta(N_{4\times 4}) \leqslant 8$, which means that $\beta(N_{4\times 4}) = 8$. Now, for $n > 4$ and n even, we can use Schwenk's Theorem, which tells us that there is a knight's tour of the $n \times n$ chessboard, and so Greenberg's argument applies. And, since half the squares are white, and half the squares are black, $\beta(N_{n\times n}) = \frac{1}{2}n^2$.

The odd case we will split into two subcases. For n of the form $4k + 1$, we can use Problem 3.3, which told us that a $(4k + 1) \times (4k + 1)$ chessboard has an open knight's tour, as was shown in Figure 3.10. So Greenberg's argument can be applied to this subcase as well, and again the best you can do is alternate knights along the tour. And, since there is one extra white square, $\beta(N_{n\times n}) = \frac{1}{2}(n^2 + 1)$.

Finally, for n of the form $4k + 3$, we will use the blocks of squares shown in Figure 10.12, in which we have labeled pairs of squares in each block in such a way that only one square of each pair can be occupied by an independent knight. Thus, the 2×4 rectangle can contain at most 4 independent knights, the 3×3 square rectangle can contain at most 5 independent knights, and the 3×4 rectangle can contain at most 6 independent knights. Since any $(4k + 3) \times (4k + 3)$ chessboard can easily be divided into rectangles of these dimensions our result follows, as illustrated in Figure 10.12 for an 11×11 chessboard, which therefore can contain at most 61 independent knights. In general, in this way, we can have at most $4(2k^2) + 6(2k) + 5$ independent knights on a $(4k + 3) \times (4k + 3)$ chessboard, and this reduces to $\frac{1}{2}((4k + 3)^2 + 1)$, as expected. This completes the proof. □

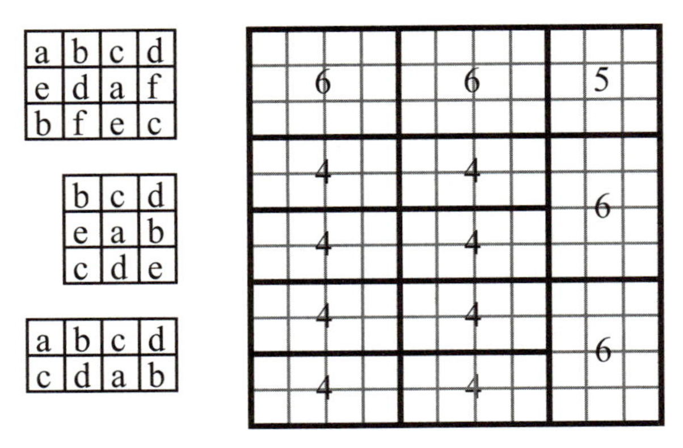

Figure 10.12 Dividing the $(4k + 3) \times (4k + 3)$ chessboard into blocks.

INDEPENDENT BISHOPS

In the previous two sections we were able not only to provide formulas for the independence numbers for the rook and for the knight, but also to see exactly how to go about producing all possible independent arrangements of those two chess pieces for any $n \times n$ chessboard. We now continue that string of successes by accomplishing the same feat for the bishop.

The first step, finding a formula for $\beta(B_{n\times n})$, is quite easy.

Theorem 10.3 $\beta(B_{n\times n}) = 2n - 2.$

Proof. As illustrated in Figure 10.13, an $n \times n$ chessboard has $2n - 1$ positive diagonals and, hence, can have at most $2n - 2$ independent bishops, since there can be at most one bishop per diagonal, and because there can't be bishops in both the upper left-hand corner and the lower right-hand corner—that is, one of those two single-square diagonals must be empty. On the other hand, you can place $n - 1$ bishops along the top row, and $n - 1$ bishops along the bottom row, as shown in Figure 10.13. Since these $2n - 2$ bishops are independent, it follows that $\beta(B_{n\times n}) = 2n - 2$. □

Now, as we approach the harder question—the most general form of the n-bishops problem—of deciding in exactly how many different ways we can arrange $2n - 2$ independent bishops on an $n \times n$ chessboard, you might recall that at the very

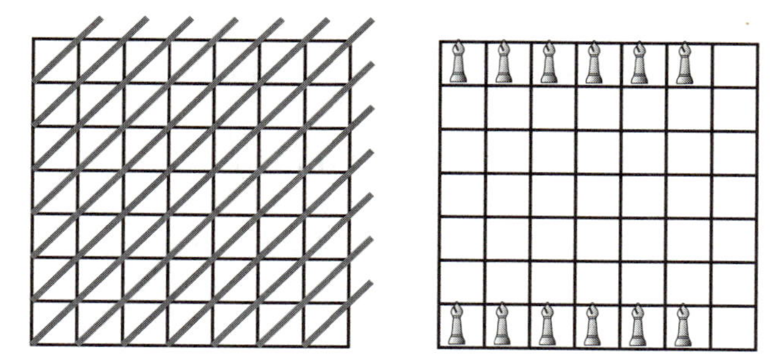

Figure 10.13 $\beta(B_{n\times n}) = 2n - 2$.

beginning of this chapter in Figure 10.1 we could easily see from the bishops graph that a maximum independent set of bishops could be achieved for the 4×4 chessboard *only* by placing bishops along the outer ring of edge squares on the board. This, of course, is also true of our placement of the $2n - 2$ bishops in Figure 10.13. This makes a certain amount of sense for any board, since it is easily seen that bishops on the outer ring of edge squares on any chessboard control fewer squares than bishops placed nearer the center of the board. A number of writers, Ball, Dudeney, and Madachy, may have made this the basis of concluding that this placement along the outer ring of edge squares was necessary, but Yaglom and Yaglom appear to have been the first to give a solid argument [40].

Theorem 10.4 (Yaglom and Yaglom) *In any arrangement of $2n - 2$ independent bishops on an $n \times n$ chessboard, all of the bishops lie on the outer ring of squares.*

Proof. Assume that $2n - 2$ independent bishops have been arranged on the board. Label each square of the board with the number of bishops that control that particular square. Clearly, a square can't be controlled by more than two independent bishops, nor can a square have a label of 0, since we could then place a new independent bishop on such a square. So, each square on the board is labeled either 1 or 2.

Our first claim is that there are at least $2n$ squares labeled 1: the 4 corner squares are clearly all labeled 1, whether they are occupied or not, and, of course, at least 2 of the corners must be unoccupied; the $2n - 2$ squares with bishops on them are also each labeled 1; together this makes at least $2n$ squares that are labeled 1—namely, the $2n - 2$ occupied squares, plus at least 2 unoccupied corner squares.

Let S be the sum of the labels for the entire chessboard. Then, since there are $2n$ squares that have a label of 1, and another $n^2 - 2n$ squares that have a label of 2 or less:

$$S \leqslant (1)(2n) + (2)(n^2 - 2n) = (2n - 2)n.$$

A bishop in the outer ring of edge squares controls exactly n squares; whereas, a bishop in the interior controls at least $n + 2$ squares—in fact, this number goes up by $+2$ for every one of the nested rings in which a bishop might be placed as it is moved toward the center of the board, exactly as was the case in Problem 10.1 for the number of squares diagonally under attack by a queen. Let a be the number of bishops in the interior of the board, and let b be the number of bishops in the outer ring. Since, by controlling a square, a bishop contributes 1 to its label, and since $a + b = 2n - 2$, we conclude that

$$S \geqslant bn + a(n + 2) = (a + b)n + 2a = (2n - 2)n + 2a.$$

Comparing this with the previous inequality for S shows that $a = 0$, which is to say that *all* bishops must be in the outer ring, as claimed. This completes the proof. □

In *Amusements in Mathematics*, Dudeney provides us with the following 'exceedingly simple rule' for determining the total number of solutions to the n-bishops problem [10].

Theorem 10.5 *There are 2^n possible arrangements of $2n - 2$ independent bishops on an $n \times n$ chessboard.*

Proof. We claim that an arrangement of $2n - 2$ independent bishops is entirely determined by what happens on the top row. Then, since each square in the top row can either contain a bishop or not, the result will follow immediately. As for the claim, whether a corner square in the top row is occupied determines whether or not the opposite corner square in the bottom row is occupied or not; similarly, for a non-corner square in

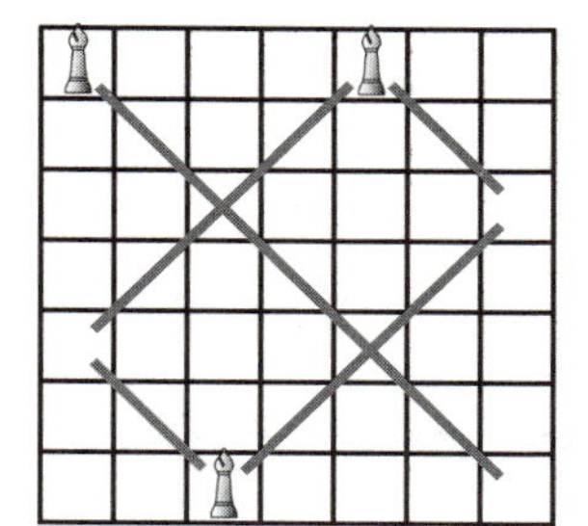
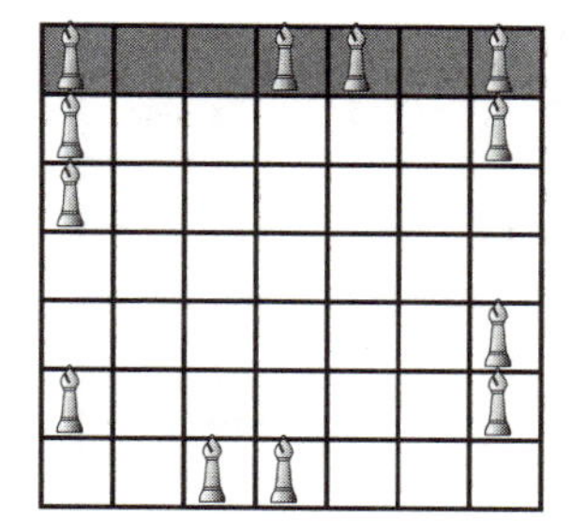

Figure 10.14 The top row determines the arrangement of independent bishops.

the top row, whether it is occupied determines what happens at the two corresponding squares in the side columns and also at the corresponding square in the bottom row, as shown on the left in Figure 10.14; in other words, if the top square has a bishop, then the side squares are empty and the bottom square must have a bishop, and vice versa if the top square is empty. An example is shown in Figure 10.14 for the 7×7 chessboard where the top row has been arbitrarily filled with four bishops, from which the rest of the arrangement follows. This completes the proof. □

In Chapter 13, we will return to this problem and determine how many of these solutions are actually different, that is, not fundamentally equivalent due to symmetry. For example, of the $256 = 2^8$ ways to arrange 14 independent bishops on an 8×8 chessboard, there are only 36 arrangements that are fundamentally different.

Independent Kings

The independence number for kings is quite easy to determine and, once again, the result is due to the Yaglom brothers [40].

Theorem 10.6 (Yaglom and Yaglom) $\beta(K_{n\times n}) = \left\lfloor \frac{1}{2}(n+1) \right\rfloor^2$.

Proof. The simple, but key, idea is that a 2×2 square may contain at most one independent king. When n is even, the $n \times n$ chessboard can be divided into $(\frac{1}{2}n)^2$ such squares and, since

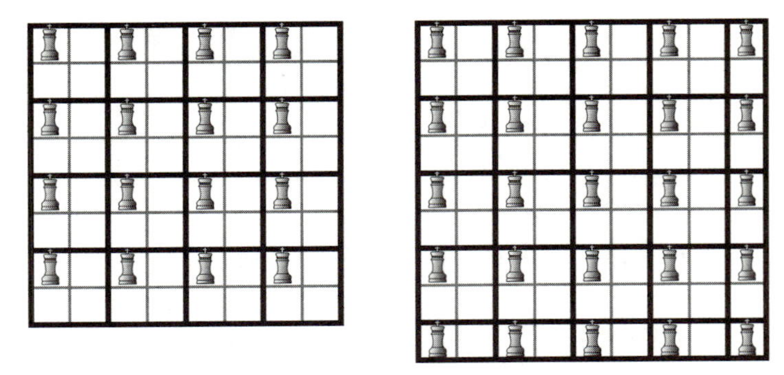

Figure 10.15 $\beta(K_{n\times n}) = \lfloor \frac{1}{2}(n+1) \rfloor^2$.

a king can be placed in, say, the upper left-hand corner of each of these 2×2 squares, the independence number in this case is given by $\beta(K_{n\times n}) = (\frac{1}{2}n)^2$. This is illustrated for an 8×8 chessboard in Figure 10.15.

The situation is similar when n is odd. As illustrated in Figure 10.15 for the 9×9 chessboard, in the odd case you can divide the board into $(\frac{1}{2}(n-1))^2$ squares of size 2×2, $n-1$ rectangles of size 2×1, and one square of size 1×1. As before, each of these regions may contain at most one independent king and, as the example in Figure 10.15 makes clear, since a king may be appropriately placed in each of these regions, the independence number in this case is given by $(\frac{1}{2}(n-1))^2 + (n-1) + 1$. This reduces to $\beta(K_{n\times n}) = (\frac{1}{2}(n+1))^2$ for n odd.

Finally, the floor function can be used to combine the two cases into a single formula. This completes the proof. □

Note that, in the odd case, the arrangement of kings is unique since there must be a king placed in the 1×1 square and this, in turn, forces the placement of all the other kings.

Solutions to Problems

Solution 10.1 This is simply a matter of figuring out, once the first queen has been placed on the board, how many of the 63 remaining squares are not under attack by the first queen. The only difficulty is that this depends somewhat on where the first queen happens to have been placed. It is true that, no matter where the first queen has been placed, it will have under attack a total of 14 squares along the row and column in which it lies. But, exactly how many additional squares it has under attack diagonally depends on how far it is from the edge of the board.

If the first queen is in the outer ring of edge squares, then this queen has 7 squares under attack diagonally, making a total of 21 squares that it has under attack. So, in this case, there are 42 squares of the 63 available squares where the second queen could be placed so that the two queens would be independent. Moreover, there are 28 of the 64 squares that are in the outer ring where the first queen could be placed. Thus, the probability that we get two independent queens by placing the first queen in the outer ring is $\frac{28}{64}\frac{42}{63}$.

If the first queen is in the next ring of squares immediately inside the outer ring, then this queen has 9 squares under attack diagonally, making a total of 23 squares that it has under attack. Thus, since there are 20 squares in this ring, the probability that we get two independent queens in this precise way is $\frac{20}{64}\frac{40}{63}$.

The next ring has 12 squares and a queen in this ring has 11 squares under attack diagonally, a total of 25 squares in all; and the final ring consists of just the 4 center squares and a queen in this ring has 13 squares under attack diagonally, a total of 27 squares in all. Thus, the probability that two randomly placed queens are independent is

$$\frac{28}{64}\frac{42}{63}+\frac{20}{64}\frac{40}{63}+\frac{12}{64}\frac{38}{63}+\frac{4}{64}\frac{36}{63}=\frac{2576}{4032}\approx .639$$

Solution 10.2 First, list all 120 permutations of the five numbers 1 through 5. Most of these permutations can be eliminated immediately, such as 1 2 5 4 3 and 2 4 3 5 1, since they contain two consecutive integers. This leaves only 14 permutations to check further, and then 4 of these—namely, 2 4 1 5 3,

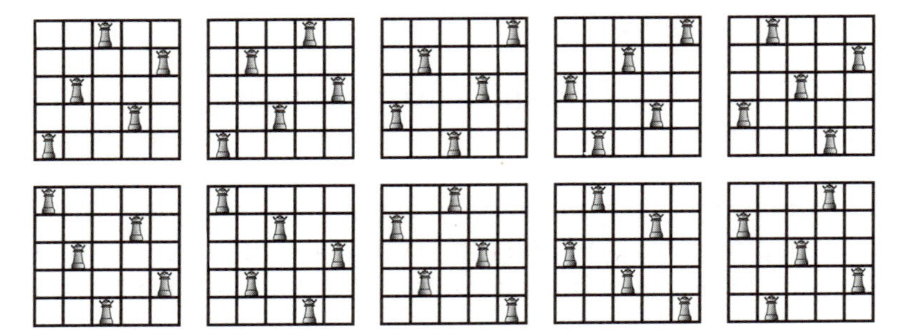

Figure 10.16 The ten solutions to the 5-queens problem.

3 1 5 2 4, 3 5 1 4 2, and 4 2 5 1 3—fail to pass Gauss's test of adding 1 2 3 4 5 and 5 4 3 2 1, since they each contain numbers that differ by 2 while being two spaces apart. Thus, there are ten solutions:

1 3 5 2 4, 1 4 2 5 3, 2 4 1 3 5, 2 5 3 1 4, 3 1 4 2 5,
3 5 2 4 1, 4 1 3 5 2, 4 2 5 3 1, 5 2 4 1 3, 5 3 1 4 2.

Eight of these are essentially the same, since they are just rotations and reflections of 1 3 5 2 4, while the other two, 2 5 3 1 4 and 4 1 3 5 2, are invariant under rotations, but are the mirror image of one another. These ten solutions are shown in Figure 10.16. Note that each of these solutions can be generated by a repeated knight's move on a torus.

Solution 10.3 By grouping the 16 queens into four 2×2 clumps, you can use an essentially unique solution to the 4-queens problem to get the first solution shown in Figure 10.17. Dudeney's solution to the second, more restricted, problem appears on the right in Figure 10.17.

Solution 10.4 As you can see in Figure 10.18, in each case, the queen placed in the bottom row falls on a diagonal already occupied by a queen previously placed in the left half of the board.

Solution 10.5 A rook in the jth row down and the kth column occupies the square with number $8(j-1)+k$. Since 8 independent rooks will necessarily collectively attain all 8 different values of k, and all 8 different values of j, we can sum these values

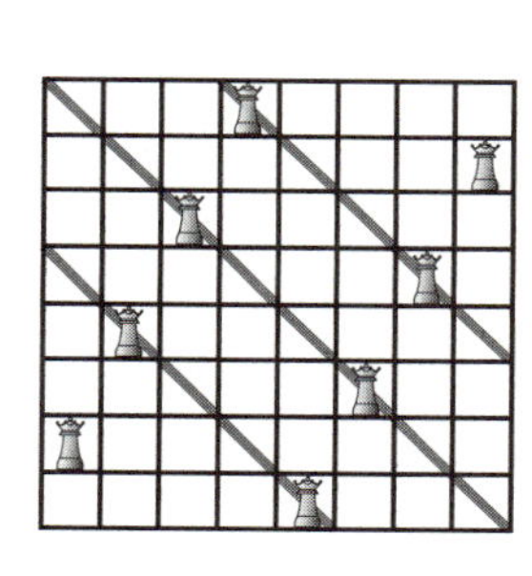

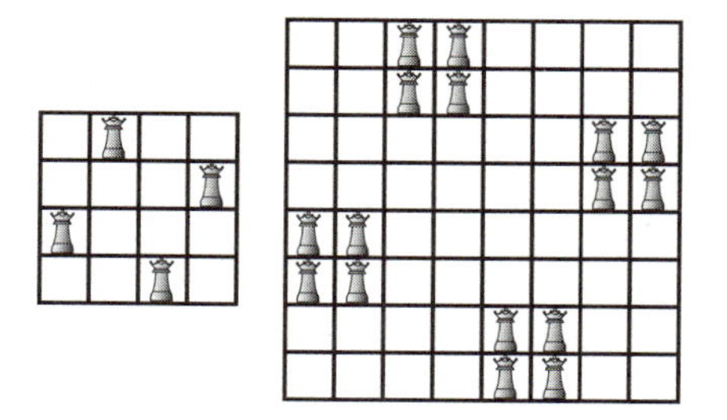

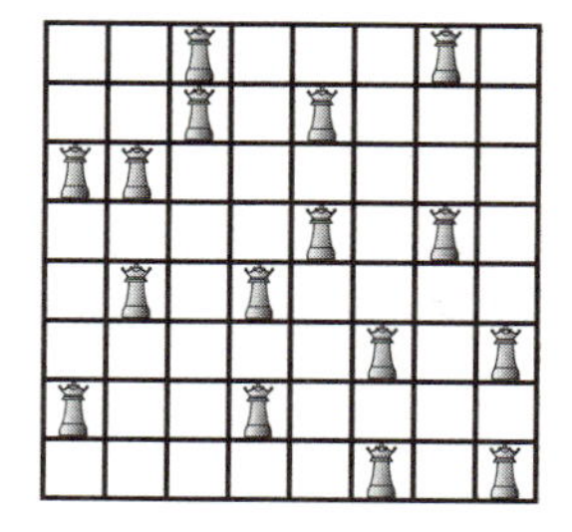

Figure 10.17 Sixteen queens without three queens in a straight line.

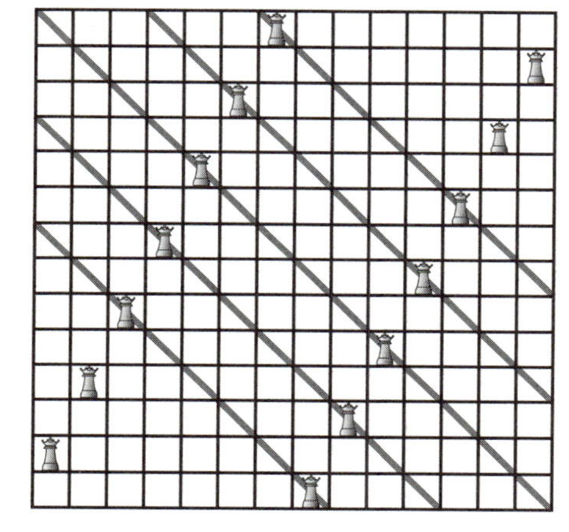

Figure 10.18 The basic construction fails for $n = 8$ and $n = 14$.

separately—irrespective of where the rooks actually are—and get

$$(0 + 8 + 16 + \cdots + 56) + (1 + 2 + 3 + \cdots + 8) = 260.$$

Note that this sum, 260, is also the magic constant for 8×8 magic squares.

CHAPTER ELEVEN

Other Surfaces, Other Variations

In this chapter, we will look to see what happens to the independence number for each of the chess pieces on a variety of other surfaces, such as the torus and the Klein bottle, and we will also come to see that, in addition to the two numbers on which we have been focusing our attention—the domination number and the independence number—there are several other interesting numbers related to chessboard problems that are natural to investigate as well. Let's start by looking at the independence problem for each of the chess pieces on the $2 \times 2 \times 2$ cube.

Problem 11.1 How many of each chess piece can you place independently on a $2 \times 2 \times 2$ cube? That is, how many independent knights? How many bishops? How many rooks? How many queens? How many kings?

The 8-Queens Problem on a Cylinder

If we take the two solutions to the 8-queens problem shown in Figures 10.2 and 10.4 and place them on cylinders as shown in Figure 11.1—where the identification is made between the top edge and the bottom edge of the chessboard—we see that the eight queens are no longer independent since, in each case, there are now two queens occupying a positive diagonal and also, by the way, two queens occupying a negative diagonal. Note that this would remain true if the identification was made between the left and right sides of the chessboard.

So one begins to wonder if it is possible to place eight independent queens on a cylindrical 8×8 chessboard. This is an interesting question, but easy to answer because all you

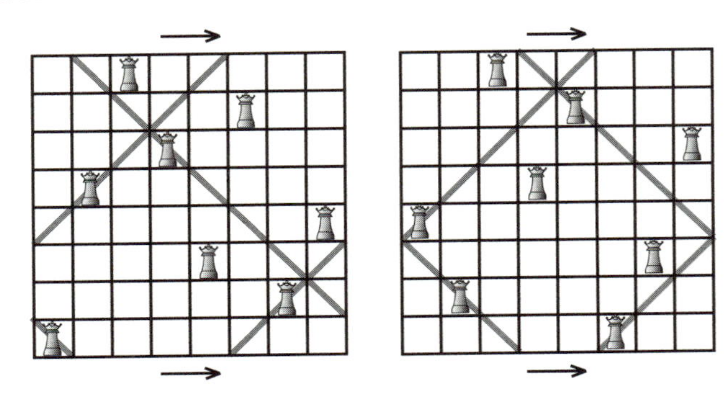

Figure 11.1 Eight queens on cylindrical 8×8 chessboards.

have to do is try each of the twelve fundamental solutions of the 8-queens problem for the regular 8×8 chessboard. It turns out that none of them works on a cylinder. Miodrag Petković gives a very nice alternative proof of this in [24], a proof which is due to E. Gik. Let us number the negative toroidal diagonals of the chessboard using the numbers 8 through 1 beginning with the negative toroidal diagonal that passes through square $(8, 8)$ and working toward the right. Yes, to the right! Then, label each square (i, j) of the chessboard with an ordered triple (i, j, k), where k is the number of the diagonal passing through square (i, j). This labeling is shown in Figure 11.2.

Note that this particular numbering of the diagonals was chosen precisely so that for each square (i, j, k) on the chessboard the sum $i + j + k$ would be divisible by 8. But, if we now have eight independent queens placed on this cylindrical 8×8 chessboard, then the i coordinates must all be distinct, and hence the eight i coordinates for these queens sum to $1 + 2 + \cdots + 8 = 36$. Similarly, for the j and k coordinates. Thus, the sum of all three coordinates for eight independent queens would be $36 + 36 + 36 = 108$, which is impossible since 108 is not divisible by 8.

In fact, 6 is the maximum number of independent queens you can place on a cylindrical 8×8 board. For example, in Figure 11.1, in each case, since there are only two pairs of doubled queens, you can remove one of the doubled queens from a pos-

1,8,7	2,8,6	3,8,5	4,8,4	5,8,3	6,8,2	7,8,1	8,8,8
1,7,8	2,7,7	3,7,6	4,7,5	5,7,4	6,7,3	7,7,2	8,7,1
1,6,1	2,6,8	3,6,7	4,6,6	5,6,5	6,6,4	7,6,3	8,6,2
1,5,2	2,5,1	3,5,8	4,5,7	5,5,6	6,5,5	7,5,4	8,5,3
1,4,3	2,4,2	3,4,1	4,4,8	5,4,7	6,4,6	7,4,5	8,4,4
1,3,4	2,3,3	3,3,2	4,3,1	5,3,8	6,3,7	7,3,6	8,3,5
1,2,5	2,2,4	3,2,3	4,2,2	5,2,1	6,2,8	7,2,7	8,2,6
1,1,6	2,1,5	3,1,4	4,1,3	5,1,2	6,1,1	7,1,8	8,1,7

Figure 11.2 Impossibility of 8 independent queens.

itive diagonal and one of the doubled queens from a negative diagonal and be left with 6 independent queens. With the exception of the symmetric solution 3 5 2 8 1 7 4 6—which can only be reduced to a set of 5 independent queens—each of the fundamental solutions for the regular 8×8 board can be reduced to a set of 6 independent queens on the cylindrical 8×8 board. Since any set of 7 queens placed independently on a cylindrical 8×8 chessboard leaves only one column unoccupied, we can assume that the first four columns each have a queen. Since the board is a cylinder, we can assume that the queen in the first column is at square $(1, 1)$. By symmetry, then, there are only three options for where to place the queen in the second column. It only takes a few minutes to check all of the possible outcomes for the first four queens and confirm that in all cases it is impossible to place 7 independent queens. Thus, $\beta(Q^{\text{cyl}}_{8\times 8}) = 6$.

On the other hand, as you can see in Figure 10.16, both of the fundamental solutions—and hence all ten solutions—to the 5-queens problem remain independent on a cylindrical board, and so $\beta(Q^{\text{cyl}}_{5\times 5}) = \beta(Q_{5\times 5}) = 5$. Similarly, both solutions shown in Figure 10.5 to the 7-queens problem remain independent on a cylindrical board, and so $\beta(Q^{\text{cyl}}_{7\times7}) = \beta(Q_{7\times7}) = 7$.

So, determining $\beta(Q^{\text{cyl}}_{n\times n})$ for an arbitrary n looks like an interesting question, and one that might not be too hard. Incidentally, the argument we used to show that $\beta(Q^{\text{cyl}}_{8\times8}) < 8$ can be used for any even n since for an $n \times n$ board the sum of the coordinates, $\frac{1}{2}(3n(n+1))$, must be divisible by n, which means that $\frac{1}{2}(3(n+1))$ is an integer, which can be the case only if n is odd.

Independent Kings on a Torus

The basic ideas used in Chapter 9 for kings domination on the torus carry over easily for independence. For an $m \times n$ chessboard any horizontal band of two rows can contain at most $\lfloor \frac{1}{2}n \rfloor$ independent kings. Since there are exactly m such bands, we conclude that there can be at most

$$\left\lfloor \frac{m\lfloor \frac{1}{2}n \rfloor}{2} \right\rfloor$$

independent kings. But we can make an identical argument for columns, so we also get

$$\left\lfloor \frac{n\lfloor \frac{1}{2}m \rfloor}{2} \right\rfloor$$

as an upper bound. As you might expect, the minimum of these two upper bounds conveniently turns out to be the kings independence number [34].

Theorem 11.1 (Watkins and Ricci) *The toroidal kings independence number for a rectangular $m \times n$ chessboard is given by*

$$\beta(K^{\text{torus}}_{m\times n}) = \min\left\{\left\lfloor \frac{m\lfloor \frac{1}{2}n \rfloor}{2} \right\rfloor, \left\lfloor \frac{n\lfloor \frac{1}{2}m \rfloor}{2} \right\rfloor\right\}.$$

Proof. If either m or n is even, it is clear that the chessboard can be divided into either horizontal bands two rows wide or vertical bands two columns wide and simply filled with the appropriate number of kings. Now assume that $m \leqslant n$ with m and n

odd. This means that of the two upper bounds

$$\left\lfloor \frac{n\lfloor \frac{1}{2}m \rfloor}{2} \right\rfloor$$

is the minimum. Writing $m = 2k + 1$, there are two cases according to whether k is even or odd.

When k is even,

$$\left\lfloor \frac{n\lfloor \frac{1}{2}m \rfloor}{2} \right\rfloor = \left\lfloor \frac{nk}{2} \right\rfloor = \frac{nk}{2},$$

and so we are trying to pack $\frac{1}{2}nk$ kings onto the chessboard. We can place $\frac{1}{2}(n-1)$ kings in each of $\frac{1}{2}(m-1)$ bands of two rows, and then we place $\frac{1}{2}k$ kings in the final row, making a total of

$$\frac{(m-1)(n-1)}{4} + \frac{k}{2} = \frac{2k(n-1)}{4} + \frac{k}{2} = \frac{kn}{2}$$

kings placed on the board. The exact placement of these kings is illustrated in Figure 11.3 for a 9×13 chessboard. In this case, $\frac{1}{2}(n-1) = 6$, and the 6 kings to be placed in each band of 2 rows can be evenly split between the two rows. If $\frac{1}{2}(n-1)$ happens to be odd, for example for a 9×11 board, the 5 kings for each band can simply be split in an alternating fashion: 3 in one row and 2 in the next for each band. Note that, since each king is placed two squares to the right of the preceding king, the total shift by the time you get to the last king is $2(\frac{1}{2}nk - 1) = nk - 2 \equiv -2 \bmod n$, and so, in the case when k is even, the last king is placed in the lower right-hand corner in the next-to-last square. Moreover, since $m \leqslant n$, all of the $\frac{1}{2}k$ kings placed in the bottom row will be independent of the kings placed in the top row.

When k is odd,

$$\left\lfloor \frac{n\lfloor \frac{1}{2}m \rfloor}{2} \right\rfloor = \left\lfloor \frac{nk}{2} \right\rfloor = \frac{nk-1}{2},$$

and so we are trying to pack $\frac{1}{2}(nk-1)$ kings onto the chessboard. This can be done in exactly the same way, but with

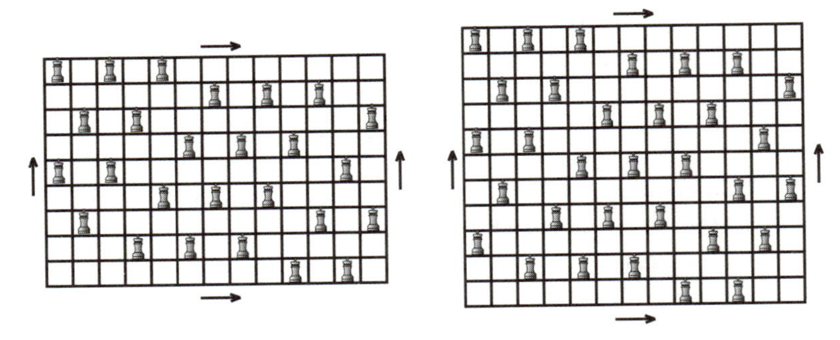

Figure 11.3 Independent kings on toroidal 9×13 and 11×13 chessboards.

$\frac{1}{2}(k-1)$ kings in the last row, and this time the total shift is $2(\frac{1}{2}(nk-1)-1) = nk-3 \equiv -3 \bmod n$. The placement of kings in this case is illustrated in Figure 11.3 for an 11×13 chessboard. □

Independent Kings on a Klein Bottle

Similarly, the basic ideas used in Chapter 9 for kings domination on the Klein bottle also carry over easily for independence. However, for independence we use panels from the top and bottom of width two that are connected, and hence make single panels of length $2n$, working our way toward the center until we are left with just one panel of width 0, 1, 2, or 3 and of length n in the middle to be dealt with. The following theorem summarizes the result [34].

Theorem 11.2 (Watkins and McVeigh) *The kings independence number for a rectangular $m \times n$ chessboard on a Klein bottle is given by*

$$\beta(K_{m\times n}^{\text{klein}}) = \begin{cases} \lfloor \frac{1}{2}m \rfloor \lfloor \frac{1}{2}n \rfloor, & \text{for } n \text{ even;} \\ n\lfloor \frac{1}{4}m \rfloor - 1, & \text{for } n \text{ odd, } m \equiv 0 \bmod 4; \\ n\lfloor \frac{1}{4}m \rfloor, & \text{for } n \text{ odd, } m \equiv 1 \bmod 4; \\ n\lfloor \frac{1}{4}m \rfloor + \lfloor \frac{1}{2}n \rfloor, & \text{for } n \text{ odd, } m \equiv 2, 3 \bmod 4. \end{cases}$$

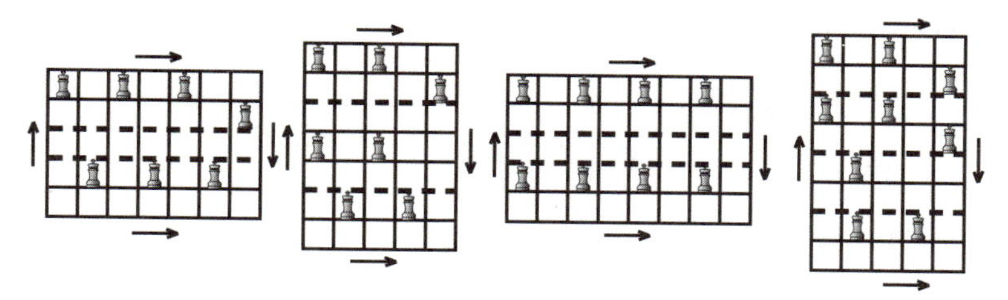

Figure 11.4 Independent kings on the Klein bottle for the 5×7, 7×5, 5×8, and 8×5 chessboards.

This result is illustrated in Figure 11.4 for each of the four cases. Note that for the 7×5 board it would seem natural to place the third king in the upper right-hand corner, but on the Klein bottle this would already not be independent of the first king in the upper left-hand corner since they would attack one another diagonally. This accounts for the shift down of the third king in the very first panel and, on this board, causes no harm. On the other hand, in general, this is what causes the -1 in the $m \equiv 0$ case in Theorem 11.2 as can be seen in the 8×5 board where there is eventually no room for a fifth king in the final panel.

Other Variations

We have concentrated in these past five chapters on domination and independence problems, but there are a number of other variations that have attracted attention as well. We now turn our attention to several of these variations, beginning with an old problem that combines the two notions of domination and independence in an obvious way.

The Independent Covering Problem

As we observed in Chapter 10, it is clear that any maximum independent set is also a dominating set; in other words, if we have placed as many independent chess pieces on a board as possible, then they must cover the entire board. On the other hand, when trying to cover a chessboard with as few pieces as possible, it seems likely that these pieces will be independent,

Figure 11.5 $\gamma(Q_{4\times4}) = 2$; $i(Q_{4\times4}) = 3$; $\beta(Q_{4\times4}) = 4$.

but they need not be. For example, in Figure 11.5 we see that two queens can cover a 4×4 chessboard, but they fail to be independent; three queens are needed to cover the board if we require them to be independent; and even four independent queens can be placed on the 4×4 chessboard.

Thus, it is natural to give this intermediate concept a name, which we of course state in terms of the underlying graph.

Definition 11.1 The *independent domination number* of a graph G is the minimum size of an independent dominating set in the graph G. We denote this independent domination number of the graph G by $i(G)$.

Note that for any graph G,

$$\gamma(G) \leqslant i(G) \leqslant \beta(G).$$

Problem 11.2 The example in Figure 11.5 seems to be somewhat of an exception. I know of only two other values of n for which $i(Q_{n\times n})$ and $\gamma(Q_{n\times n})$ differ, namely, $n = 6$ and $n = 12$. A minimum dominating set with 3 queens is shown in Figure 8.8 for the 6×6 board, and one with 6 queens is shown in Figure 8.11 for the 12×12 board. Find an independent covering with 4 queens for the 6×6 board and one with 7 queens for the 12×12 board.

It has been conjectured by Fricke and others in [11] that, except for a few small values of n, $i(Q_{n\times n}) = \gamma(Q_{n\times n})$ for all n. Lending support to this conjecture is something I didn't mention at the time, but is something you might have noticed during the lengthy proof of Theorem 8.3: that the dominating set of queens in which Spencer's lower bound of $\frac{1}{2}(n-1)$ is achieved turns out to be independent; in other words, at least in this case, $i(Q_{n\times n}) = \gamma(Q_{n\times n})$.

You might have also been aware during Chapter 8 that the Spencer–Cockayne construction actually produced independent sets of queens that dominated chessboards. Thus, they were able to use this construction to provide an upper bound for the queens independent domination number [6]: $i(Q_{n\times n}) < .705n + .895$.

Weakley has now shown that an upper bound for $i(Q_{n\times n})$ is essentially given by $\frac{43}{75}$ of n. Similarly, Burger and others have shown that an upper bound for $\gamma(Q_{n\times n})$ is essentially given by $\frac{31}{54}$ of n. Since $\frac{43}{75} = .573$, and $\frac{31}{54} = .574$, this is further evidence that $i(Q_{n\times n})$ and $\gamma(Q_{n\times n})$ may be equal except for a finite number of values of n.

There are, of course, several chess pieces for which $i = \gamma$ on the $n \times n$ chessboard for *all* values of n, namely, the king, the rook, and the bishop. In the standard lattice pattern of kings used to cover the chessboard, as shown in Figure 7.7, the kings are independent. Rooks can be placed all along a main diagonal to cover the chessboard while still being independent, as in Figure 7.4. And bishops placed along a single central column, as in Figure 7.6, are also independent. Thus, for all of these pieces, the independent domination number is the same as the domination number itself.

You may remember that the proof given in Chapter 7 for bishops that $\gamma(B_{n\times n}) = n$ was due to Yaglom and Yaglom, and involved looking at black and white squares separately, odd and even cases, rotating the board 45° and thinking of the bishops as rooks. Here is a very clean and slick combinatorial proof of the same result that is due to Cockayne, Gamble, and Shepard [6].

Theorem 11.3 (Cockayne, Gamble, and Shepard)

$$\gamma(B_{n\times n}) = i(B_{n\times n}) = n, \quad \textit{for all } n.$$

Proof. Since we have already observed that placing n bishops along a central column of the chessboard produces an independent dominating set we know that $\gamma(B_{n\times n}) \leqslant i(B_{n\times n}) \leqslant n$. So all we have to do is show that $\gamma(B_{n\times n}) \geqslant n$.

Assume that we have a minimum dominating set of bishops on the board and that this set of bishops contains n_{W} white bishops and n_b black bishops. First, number the negative diagonals of the chessboard starting at the lower left-hand corner and going all the way to the upper right-hand corner with the numbers from 1 to $2n - 1$. Next, let w be the negative white diagonal that contains no bishop that is closest to the main negative diagonal; we can assume that $w \leqslant n$, otherwise simply rotate the board about the main negative diagonal. Similarly, define b for the negative black diagonals; we can also assume that $b \leqslant n$, otherwise rotate the black squares about the main diagonal leaving all the white squares in place—this does not affect the minimum dominating set.

Because of the way in which we numbered the diagonals, diagonal w contains precisely w squares, but since diagonal w itself contains no bishop, each of these w squares must be controlled by a distinct white bishop along a positive diagonal. Thus, $n_w \geqslant w$. Similarly, $n_b \geqslant b$.

Also, by the way diagonal w was chosen, we know that there is at least one bishop on each of the negative white diagonals strictly between diagonal w and diagonal $2n - w$, but there are $n - w - 1$ such white diagonals between these two. Thus, $n_w \geqslant n - w - 1$. Similarly, $n_b \geqslant n - b - 1$.

The claim is that at least one of these four inequalities must be a strict inequality. Otherwise, if all four inequalities are in fact equalities, then $w = n_w = n - w - 1$, and so $w = \frac{1}{2}(n - 1)$; and, similarly, $b = n_b = n - b - 1$, and so $b = \frac{1}{2}(n - 1)$. But then $w = b$, which is impossible, since the diagonals of one color are always odd and the diagonals of the other color are always even on any chessboard.

Therefore, since at least one of the four inequalities is strict, we conclude that

$$n_w + n_b + n_w + n_b > w + b + (n - w - 1) + (n - b - 1) = 2(n - 1),$$

and so $n_w + n_b > n - 1$; that is, $n_w + n_b \geqslant n$, which is exactly what we wanted to show: there were at least n bishops in the minimum dominating set all along. □

Problem 11.3 It is easy to see that all of the knights coverings given in Figure 7.2 in Chapter 7, except those for the 7×7 and

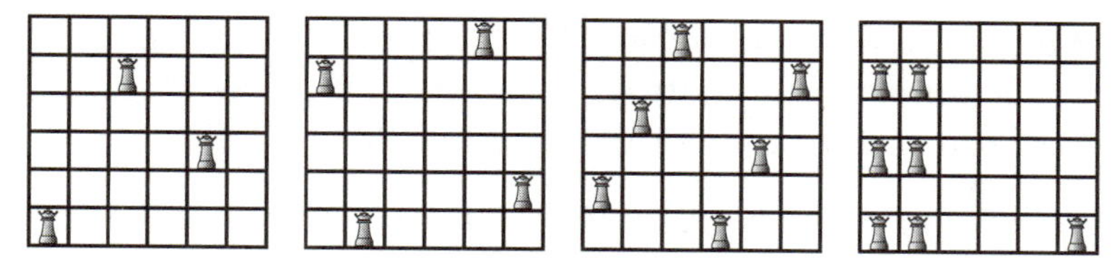

Figure 11.6 Minimal dominating sets.

the 8×8 boards, are independent. In other words, except for $n = 7$ or 8, $\gamma(N_{n\times n}) = i(N_{n\times n})$ for all $n \leqslant 10$. Find an independent dominating set with 13 knights for the 7×7 chessboard, and one with 14 knights for the 8×8 chessboard. In each case, this is the best that you can do.

Two excellent surveys, *Combinatorial Problems on Chessboards: A Brief Survey* by Fricke, Hedetniemi, Hedetniemi, McRae, Wallis, Jacobson, Martin, and Weakley [11], and *Combinatorial Problems on Chessboards: II* by Hedetniemi, Hedetniemi, and Reynolds [18] provide an extensive, detailed, and up-to-date discussion of domination, independence, and independent domination problems on chessboards. In addition, they discuss several other variations of these standard problems that deserve our attention.

The Upper Domination Number

Consider the four dominating sets of queens on the 6×6 chessboard shown in Figure 11.6. Obviously, only the set on the left is a minimum dominating set. However, each of these four sets share the following remarkable property: if you remove any one queen from any of these sets, the remaining queens will no longer cover the entire board. This is easy to confirm by simply checking each queen one at a time. The interesting question is then: how big can a set get and still have this property?

This leads us to the following definition.

Definition 11.2 A dominating set S in a graph G is said to be a *minimal dominating set* if the removal of any vertex from S results in a set that no longer dominates the graph G; that is, no proper subset of S is itself a dominating set of the graph G. The *upper domination number* of a graph G is, then, the maximum size of a minimal dominating set of the graph G. We denote this upper domination number of the graph G by $\Gamma(G)$.

We have previously observed that a maximum independent set must be a dominating set, that is, for any graph G, $\gamma(G) \leqslant \beta(G)$; moreover, it is clear that a maximum independent set must also be a minimal dominating set, that is, $\beta(G) \leqslant \Gamma(G)$. Thus, we always have the following inequalities:

$$\gamma(G) \leqslant i(G) \leqslant \beta(G) \leqslant \Gamma(G).$$

The minimal dominating set of 7 queens on the right in Figure 11.6 was found by Weakley and shows that $\beta(Q_{6\times6}) < \Gamma(Q_{6\times6})$. For smaller values of n, β and Γ coincide. Weakley has shown that for $n \geqslant 5$, $\Gamma(Q_{n\times n}) \geqslant 2n - 5$. It is now known that $\Gamma(Q_{6\times6}) = 7$ and that $\Gamma(Q_{7\times7}) = 9$, and so, for $n = 5, 6, 7$, Weakley's bound is exact, but $n = 7$ is the largest value of n for which $\Gamma(Q_{n\times n})$ is known.

Given the high degree of regularity with which we have been able to approach domination and independence problems for kings, it is somewhat surprising how little is known about the upper domination number for kings. About all we know is that β and Γ coincide for values of n up to $n = 7$. I think that by looking at the example of a minimal dominating set of 37 kings in Figure 11.7 for a 12×12 chessboard, where it is paired with a maximum independent set of 36 kings, you can begin to see that studying the upper domination number is really quite different from anything else we have encountered up to this point. This spectacular example, found by McRae using a genetic algorithm, shows that $\beta(K_{12\times12}) < \Gamma(K_{12\times12})$.

Turning now to rooks, a moment of reflection makes it clear that $\Gamma(R_{n\times n}) = n$, since a dominating set of rooks, minimal or otherwise, dominates a chessboard by either having at least one rook in every row or having at least one rook in every column; and so a dominating set of rooks can be minimal only by either having exactly one rook in every row or having exactly one rook in every column. Hence, for rooks, $\Gamma = \beta$.

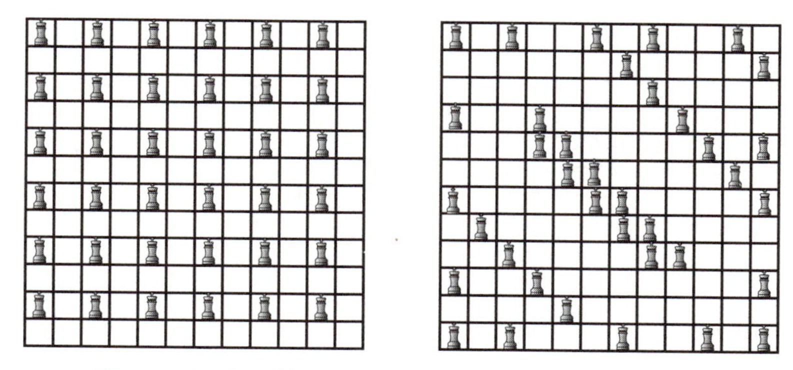

Figure 11.7 $\beta(K_{12\times 12}) = 36$ and $\Gamma(K_{12\times 12}) \geqslant 37$.

For bishops it is equally easy to see that $\Gamma = \beta$. An $n \times n$ chessboard has $2n - 1$ positive diagonals and a minimal dominating set of bishops can have at most one bishop on each of these positive diagonals, and cannot have a bishop on both of the shortest positive diagonals, that is, at both the upper left-hand corner and the lower right-hand corner. Therefore, $\Gamma(B_{n\times n}) \leqslant 2n - 2$. But, we know that $\Gamma(B_{n\times n}) \geqslant \beta(B_{n\times n}) = 2n - 2$. Therefore, $\Gamma(B_{n\times n}) = 2n - 2$.

And for knights we also know that $\Gamma = \beta$ because of a theorem due to Cockayne, Favaron, Payan, and Thomason [7] that says that whenever G is a bipartite graph, $\Gamma(G) = \beta(G)$. Recall that a graph is said to be *bipartite* if the vertices of G can be colored black and white in such a way that every edge in the graph joins a black vertex and a white vertex, and so the knights graph is quintessentially bipartite merely as a consequence of the standard checkerboard coloring of the chessboard.

The Irredundance Numbers

In trying to cover a chessboard with as *few* chess pieces as possible, it is obvious that redundancy is a bad thing. That is, we don't want to cover the board, or even a portion of it, with a given set of chess pieces when we could achieve the same coverage by removing a redundant piece. This common sense notion leads us quite naturally to a study of *irredundant* sets.

Figure 11.8 Irredundant sets.

Definition 11.3 A set S of vertices in a graph G is called *irredundant* if for each vertex v in S either v itself is not adjacent to any other vertex in S or else there is at least one vertex u not in S such that u is adjacent to v but to no other vertex in S.

This sounds more complicated than it really is. In terms of chess, it simply says that a chess piece is *not* redundant if it either occupies a square that is not covered by some other piece or else it covers some square that no other piece covers. For example, consider the three sets of kings placed on the 7×7 chessboard shown in Figure 11.8. Each of these sets is irredundant. Note, in particular, that the first set does not cover the entire chessboard, but it is still irredundant, since removing any one of the kings reduces the total number of squares covered. Note also that in the last set in Figure 11.8 removing any one of the kings reduces the total number of squares covered by exactly 1, namely, by the square occupied by the removed king.

The first set of eight kings on the left in Figure 11.8 has an interesting property: if we place an additional king anywhere on the board, then the new set of nine kings will fail to be irredundant; that is, there will be at least one of the nine kings that is now redundant. This is obvious if the new king is placed, say, on the left side of the board, and hence covers no new squares, but is much less obvious if the new king is placed so as to cover squares that the original eight kings did not cover. You should try a few placements of this kind for the new king—for example, in any of the four upper right-hand corner squares—and observe which of the original kings becomes redundant. Needless to say, the other two irredundant sets of kings in Figure 11.8 also have this same property. This leads us to the following definition.

Definition 11.4 An irredundant set S in a graph G is said to be a *maximal irredundant set* if S is not a proper subset of any irredundant set in G. The irredundance number of a graph G is, then, the minimum size of a maximal irredundant set in the graph G. We denote this irredundance number of the graph G by $ir(G)$.

In general, since any minimum dominating set is automatically a maximal irredundant set, we have that, for any graph G, $ir(G) \leqslant \gamma(G)$. Note that the maximal irredundant set of 8 kings on the left in Figure 11.8 shows that $ir(K_{7\times 7}) \leqslant 8$; but we know that $\gamma(K_{7\times 7}) = 9$, and so $ir(K_{7\times 7}) < \gamma(K_{7\times 7})$.

Very little is known about the irredundance number for the various chess pieces with the exception of two easily proved formulas for rooks and bishops: $ir(R_{n\times n}) = n$ and $ir(B_{n\times n}) = n$.

Looking once again at Figure 11.8, this time at the last set of 16 kings on the right, you might well ask yourself whether this is the largest possible irredundant set of kings for the 7×7 chessboard. This questions leads us to yet another definition.

Definition 11.5 The *upper irredundance number* of a graph G is the maximum size of an irredundant set in the graph G. We denote this upper irredundance number of the graph G by $IR(G)$.

For example, the irredundant set of 16 kings on the right in Figure 11.8 shows us that $IR(K_{7\times 7}) \geqslant 16$. And, since you may have noticed that I failed to include a fourth chessboard with yet a larger set of irredundant kings in Figure 11.8, you would be correct in inferring that $IR(K_{7\times 7}) = 16$.

In general, since a minimal dominating set is always an irredundant set, we have that for any graph G, $\Gamma(G) \leqslant IR(G)$. Thus, we can at last finally summarize the ultimate chain of inequalities that always holds for any graph for the six numbers we have defined:

$$ir(G) \leqslant \gamma(G) \leqslant i(G) \leqslant \beta(G) \leqslant \Gamma(G) \leqslant IR(G).$$

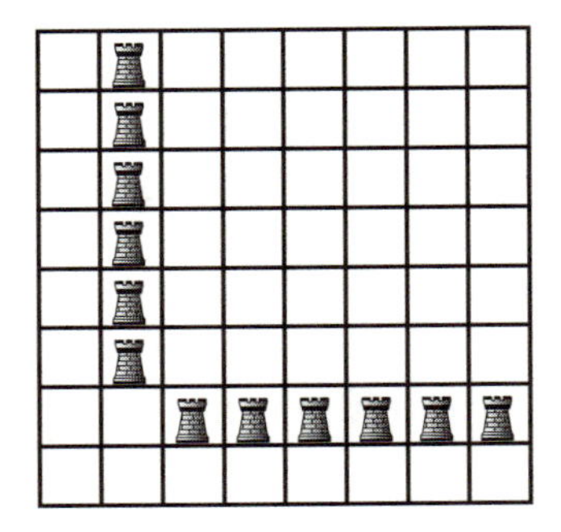

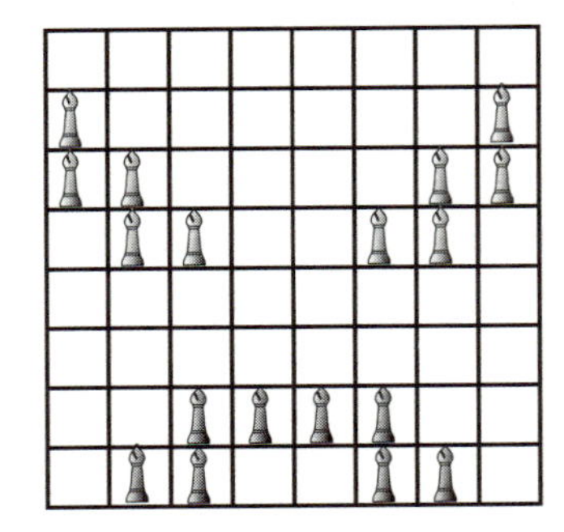

Figure 11.9 Maximum irredundant sets of rooks and bishops.

Several of the upper irredundance numbers are known [11]. For rooks, Hedetniemi, Jacobson, and Wallis have shown that for $n \geqslant 4$, $IR(R_{n\times n}) = 2n - 4$; and a maximum irredundant set of rooks is shown for the 8×8 chessboard in Figure 11.9. For bishops, Fricke has shown that for all n, $IR(B_{n\times n}) = 4n - 14$; and a maximum irredundant set of bishops for the 8×8 chessboard is shown in Figure 11.9. And for knights, we can again use the theorem of Cockayne, Favaron, Payan, and Thomason that we previously used for the knights upper domination number because, in fact, that theorem also includes the upper irredundance number, and says that whenever G is a bipartite graph, then $\beta(G) = \Gamma(G) = IR(G)$.

For kings, about all that is known is a pair of bounds due to Fricke: for $n > 1$, $\frac{1}{3}(n-1)^2 \leqslant IR(K_{n\times n}) \leqslant \frac{1}{3}n^2$. This, by the way, is why we know for sure that $IR(K_{7\times 7}) = 16$. For queens, there is little that is known in general. Maximum irredundant sets of queens are shown in Figure 11.10 for the 5×5, 6×6, 7×7, and 8×8 chessboards.

Problem 11.4 For each of the irredundant queens in Figure 11.10, find a square on the chessboard that is controlled only by that queen.

The Total Domination Number

In Problem 7.2 in Chapter 7 you were asked to solve a problem due to H. E. Dudeney, and place 14 knights on a chessboard so that every square on the board is actually attacked by a knight,

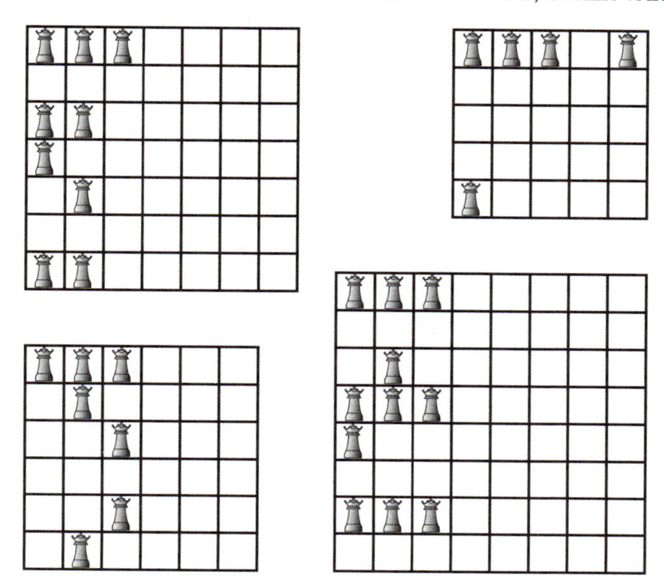

Figure 11.10 $IR(Q_{5\times5}) = 5$; $IR(Q_{6\times6}) = 7$; $IR(Q_{7\times7}) = 9$; $IR(Q_{8\times8}) = 11$.

rather than just being either occupied or attacked by a knight, as has been our usual convention in domination problems. The minimum number of chess pieces that this can be done with has now become known, for each chess piece, as the *total domination number.* The total domination problem was formally introduced by W. W. Rouse Ball [2], who gave the total domination numbers for the 8×8 chessboard for queens to be 5, for bishops 10, for knights 14, and for rooks 8.

Problem 11.5 Find an arrangement of 5 queens on the 8×8 chessboard that totally dominates the board. Do the same for 10 bishops, 14 knights, and 8 rooks.

If you have been at all intrigued by this brief discussion of 'other variations', I strongly urge you to look at the two *Combinatorial Problems on Chessboards* papers [11, 18]. They contain a wealth of interesting material, and provide an extraordinarily rich source of problems that will keep all of us busy for many years to come. What's the mathematical equivalent of *bon appétit*?

SOLUTIONS TO PROBLEMS

Solution 11.1 Since a knight, placed say on top of the $2 \times 2 \times 2$ cube, only attacks squares on one of the four sides of the cube, it is possible to place 8 independent knights on the $2 \times 2 \times 2$ cube by placing 4 knights on top of the cube and 4 more knights on the bottom of the cube, as shown in Figure 11.11. Moreover, this is the best possible. A knight placed on the top of the $2 \times 2 \times 2$ cube leaves 6 squares on the sides uncovered. It is easy to check that placing a second knight at any of these uncovered side squares then leaves only a total of either 5 or 6 uncovered squares anywhere on the cube. This is illustrated in Figure 11.11, where by symmetry we need only check three possible placements for the second knight. Thus, you can't hope to place more than 8 independent knights on the $2 \times 2 \times 2$ cube.

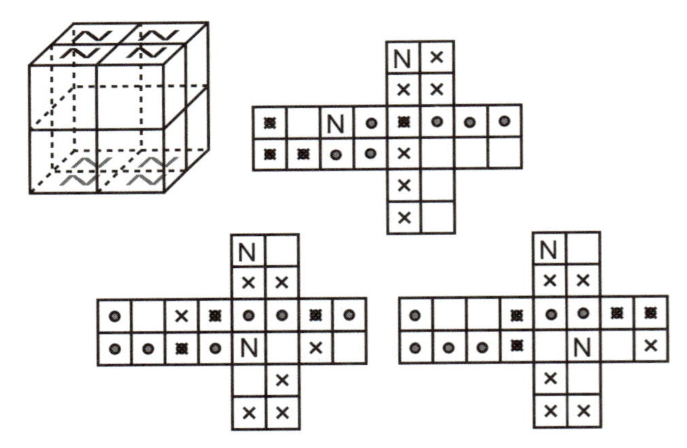

Figure 11.11 Independent knights on a $2 \times 2 \times 2$ cube.

Rather surprisingly, the maximum number of independent bishops you can place on a $2 \times 2 \times 2$ cube is 4. In Figure 11.12, I have colored the squares of the $2 \times 2 \times 2$ cube using four colors in such a way that a bishop placed on any square will automatically cover all of the other squares having that same color. Therefore, at most 4 independent bishops can be placed on the board. And a solution with 4 independent bishops is also shown in Figure 11.12.

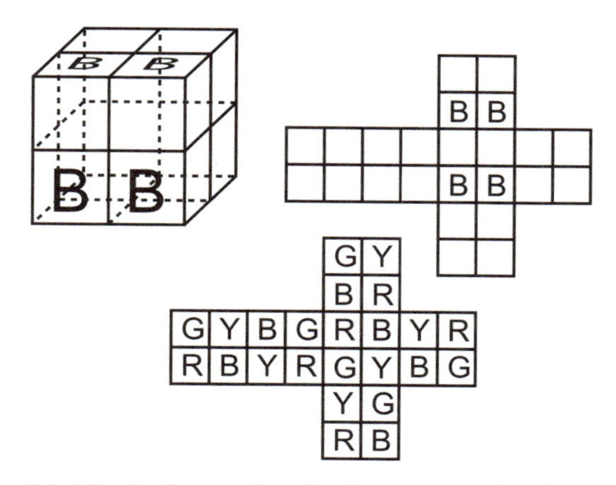

Figure 11.12 Independent bishops on a $2 \times 2 \times 2$ cube.

Note that, for each color, the colored squares correspond to the hexagonal cross-section of the cube we discussed in our solution to Problem 1.3 in Chapter 1; and a cube has exactly four of these hexagonal cross-sections because on each face of the cube the cross-section joins the midpoints of two adjacent edges, and on each face there are of course exactly four ways to join adjacent midpoints. Note also that the arrangement of colors on the six faces of the cube in Figure 11.12 represents the six different ways one could put four colors in order in a clockwise fashion.

The maximum number of independent rooks you can place on a $2 \times 2 \times 2$ cube is 3. Imagine a rook placed, say, on the front of the cube. This rook covers all of the squares horizontally in whichever of the two layers of the cube the rook is in, that is, all of the squares in a horizontal row going all the way around the cube; this rook also covers all of the squares in a vertical column going all the way around the cube, either the left or right such column. This is illustrated in Figure 11.13. Now, of course, the $2 \times 2 \times 2$ cube has six such bands of squares going around it, two horizontal bands, and two in each of the two vertical planes. Therefore, at most 3 independent rooks could be placed on the $2 \times 2 \times 2$ cube since each rook fully uses up two of these six bands. One arrangement of 3 independent rooks is shown in Figure 11.13.

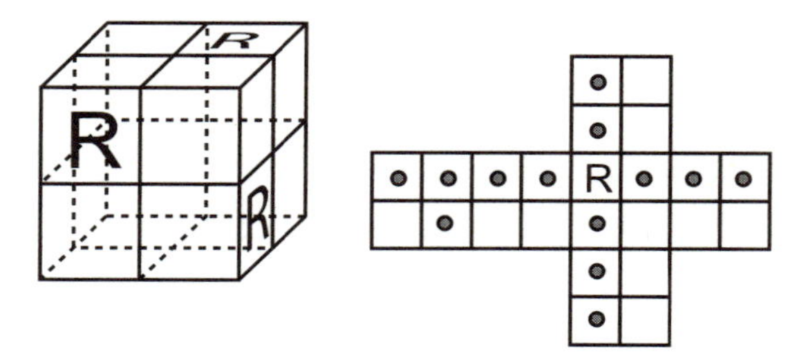

Figure 11.13 Independent rooks on a $2 \times 2 \times 2$ cube.

Since you could never place more independent queens than rooks on any chessboard, the arrangement of 3 independent queens given in Figure 11.14 then shows that the maximum number of independent queens you can place on a $2 \times 2 \times 2$ cube is 3.

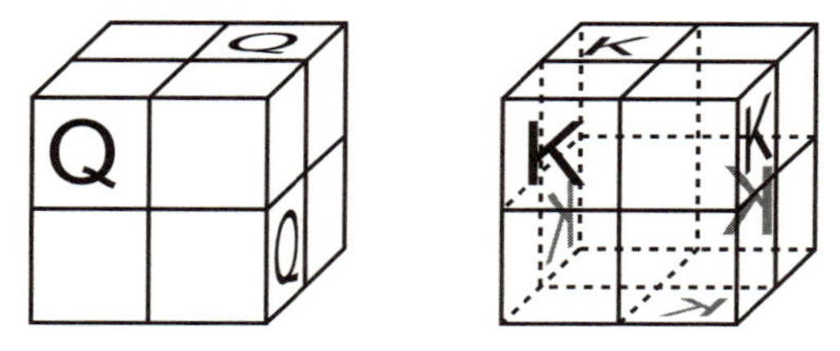

Figure 11.14 Independent queens and kings on a $2 \times 2 \times 2$ cube.

Since you could never place more than one independent king on each face of the cube, no more than 6 independent kings could be placed on the $2 \times 2 \times 2$ cube. The arrangement of 6 independent kings given in Figure 11.14 then shows that the maximum number of independent kings you can place on a $2 \times 2 \times 2$ cube is 6.

Solution 11.2 Minimum independent coverings are shown in Figure 11.15.

Solution 11.3 Two minimum independent dominating sets are shown in Figure 11.16.

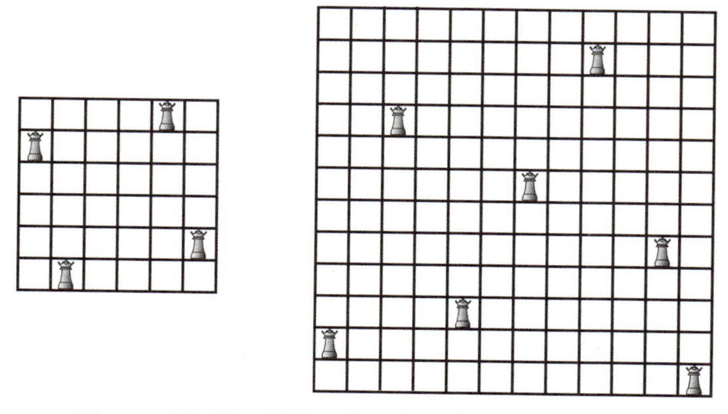

Figure 11.15 $i(Q_{6\times 6}) = 4$ and $i(Q_{12\times 12}) = 7$.

Figure 11.16 $i(N_{7\times 7}) = 13$ and $i(N_{8\times 8}) = 14$.

Solution 11.4 The shaded squares in Figure 11.17 indicate all of the squares that are controlled by only a single queen.

Solution 11.5 The original solutions given by Ball [2] are shown in Figure 11.18. Additional solutions for the knight are given in Figure 7.10.

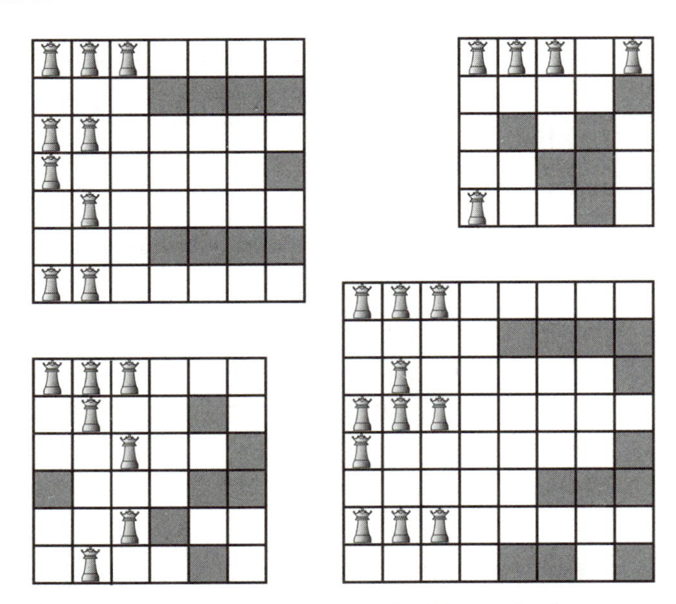

Figure 11.17 Squares controlled by a single queen.

Figure 11.18 Total domination numbers for the 8×8 chessboard.

CHAPTER TWELVE

Eulerian Squares

W. W. Rouse Ball concluded his essay on chessboard recreations in [2] with a brief discussion of Euler's contributions to the study of Latin squares. A *Latin square* is an $n \times n$ array of the integers $0, 1, 2, \ldots, n-1$—or, equivalently, if you prefer, a labeling of the squares of an $n \times n$ chessboard with these integers—such that each integer appears once and only once in each row, and once and only once in each column. There is nothing special about using integers in Latin squares, except that on occasion their arithmetical properties may be involved, and so other symbols are frequently used as well, as we shall see later. In fact, these arrays now have the rather bizarre name *Latin squares* only because of the simple happenstance that Euler chose to use Latin letters as labels in his own study of these 'quarres magiques' and because we now need a suitable name to distinguish this concept from that of 'magic squares', which, while similar, are something else entirely. Two 5×5 Latin squares are shown in Figure 12.1.

Constructing Latin Squares

It is not at all difficult to build Latin squares of any size, and W. W. Rouse Ball gives one method that is especially appealing. To construct an $n \times n$ Latin square take two integers a and b, each of which is relatively prime to n—that is, they share no common factors with n. Then, starting at square $(1, 1)$ in the lower left-hand corner, fill the bottom row with the numbers $0, b, 2b, 3b, \ldots,$ reducing modulo n as necessary; in practice, it is probably easier to think of 'adding b' modulo n at each step. For example, in the Latin square on the

4	0	3	2	1
3	2	4	1	0
2	4	1	0	3
1	3	0	4	2
0	1	2	3	4

4	1	3	0	2
3	0	2	4	1
2	4	1	3	0
1	3	0	2	4
0	2	4	1	3

Figure 12.1 Two 5×5 Latin squares.

right in Figure 12.1, we are taking $b = 2$, and so the bottom row is $0, 2, 2 + 2 = 4, 4 + 2 = 6 \equiv 1 \bmod 5$, and $1 + 2 = 3$. Next, the left-hand column is formed in the same way using the number a, and filling the column starting at $(1, 1)$ with $0, a, 2a, 3a, \ldots$, again reducing modulo n. In the example on the right in Figure 12.1, $a = 1$. Finally, the rest of the entries in the square array are determined as in an addition table using the numbers from the left column and bottom row and doing the arithmetic modulo n. Thus, for example, in the Latin square on the right in Figure 12.1, the entry in square $(4, 4)$ is 4 because $4 = 3 + 1$, and the entry in square $(3, 3)$ is 1 because $1 \equiv 2 + 4 \bmod 5$.

Problem 12.1 Construct an 8×8 Latin square using Ball's method with $a = 1$ and $b = 3$.

Eulerian Squares

We can superimpose two Latin squares of the same size by combining their entries and forming ordered pairs. If the resulting combination is such that each of the possible n^2 ordered pairs occurs exactly once, then the two Latin squares are said to be *orthogonal*, and the new combined square is called an *Eulerian square*, or sometimes the term *Graeco-Latin square* is also used because Latin letters might be used for one square and Greek letters for the other square. Note that the two Latin squares in Figure 12.1 are *not* orthogonal because the combined square has repetitions of certain pairs, such as $(0, 0)$ and $(1, 1)$, while other pairs, such as $(1, 0)$ and $(2, 3)$, are simply missing. On the other hand, the two 4×4 Latin squares shown in Figure 12.2 *are* orthogonal, and the resulting Eulerian—or Graeco-Latin—

3	2	1	0
0	1	2	3
2	3	0	1
1	0	3	2

3	2	1	0
1	0	3	2
0	1	2	3
2	3	0	1

(3,3)	(2,2)	(1,1)	(0,0)
(0,1)	(1,0)	(2,3)	(3,2)
(2,0)	(3,1)	(0,2)	(1,3)
(1,2)	(0,3)	(3,0)	(2,1)

Figure 12.2 Two orthogonal Latin squares and the combined Eulerian square.

square is shown there as well. This particular Eulerian square is in fact the solution to Ozanam's Problem, mentioned in Chapter 1 as Problem 1.6, in which you were asked to place the four aces, kings, queens, and jacks from a single deck of cards into a 4×4 array so that each row and each column of the array contained a card of each rank as well as a card of each suit.

Problem 12.2 Show that the Eulerian square in Figure 12.2 in fact yields a magic square if you think of each of the ordered pairs as a two-digit number written in base 4. For example, think of the ordered pair $(3, 2)$ as 32 base 4, which then becomes $3 \times 4 + 2 = 14$.

Euler was looking at orthogonal Latin squares in the first place precisely to produce 'quarres magiques' such as the one in Problem 12.2. Now, in retrospect, what is particularly interesting about all this is that, although Euler himself described his investigations on this matter as being 'of little use', in the 1920s it was realized that Latin squares could be very useful indeed in the design of statistical experiments in agriculture. A Latin square is an absolutely perfect design for an experiment that seeks to test different treatments of a particular crop by equalizing any possible local effects due to things such as varying soil fertility, amount of moisture received, and other uncontrollable factors. Euler would be amazed, but these days, Latin squares are widely used in many applied areas ranging from the design of experiments in medicine to product testing in business. Moreover, Eulerian, or Graeco-Latin, squares are an especially useful form of Latin squares since they automatically allow you to include an additional variable in any experiment.

This now gives us a down-to-earth practical reason to look for orthogonal Latin squares to add to what started out long ago as just a matter of natural curiosity on the part of Euler.

Euler's Conjecture

Euler knew that it was easy to construct a pair of orthogonal Latin squares whenever n is odd. All you have to do is construct one square using the method described by Ball mentioned above letting $a = 1$ and $b = 1$, and then construct another square in the same way letting $a = 1$ and $b = 2$. It is easy to show that these squares will always be orthogonal when n is odd.

On the other hand, if, for example, $n = 2$, then clearly there is no *pair* of orthogonal Latin squares simply because there are only two 2×2 Latin squares, and they obviously aren't orthogonal. But, if n is a power of 2 greater than 2, then as Ball indicated we can still use the same method as above to produce two orthogonal Latin squares merely by replacing the integers $0, 1, 2, \ldots, n-1$ by what we now call the *Galois field* $GF(n)$. In fact, for $n = 4$, this is the way I constructed the two orthogonal Latin squares in Figure 12.2—although I changed the labels of the finished product into the more familiar integers 0, 1, 2, and 3 so that the connection to Ozanam's Problem would be more obvious.

As Ball also points out, if you have a $j \times j$ Eulerian square and a $k \times k$ Eulerian square, then it is easy to construct a $jk \times jk$ Eulerian square. Therefore, the only values of n where you might not be able to construct a pair of orthogonal Latin squares would be $n = 2, 6, 10, 14, 18, \ldots$. Since Euler seemed to be aware of this and it was he who first focused attention on this particular infinite family of Latin squares, it is certainly appropriate to name Eulerian squares in his honor.

Nonetheless, Euler was unable to find a pair of orthogonal 6×6 Latin squares. As described by Martin Gardner in *New Mathematical Diversions* [15], Euler put the question this way: each of 6 different regiments has 6 officers, and each officer has one of 6 different ranks; can these 36 officers be arranged in a square formation so that each row and each file contains one officer of each rank and one of each regiment? This has become known as *Euler's 36 Officer Problem*. Euler posed this problem in 1782 shortly before his death, but it wasn't until 1900 that G. Tarry proved that no solution was possible. In a testament to the spirit of human optimism a 'solution' to the 36 Officer Problem was recently published on the Internet. But, within just three days, John Conway found a mistake—an act which in turn is a testament to the speed of communications in the current mathematical world—and Tarry's proof seems secure once again.

Euler firmly believed it to be impossible to produce such a 6×6 square, and in fact that it was similarly impossible to produce such a 10×10 square, or such a 14×14 square, or such an 18×18 square, and so on. This bold prediction—really based on little more than the fact that it was true for 2×2 Latin squares and that he was unable to produce a pair of orthogonal 6×6 Latin squares—became known as *Euler's Conjecture*: there does not exist a pair of orthogonal $n \times n$ Latin squares for $n = 2, 6, 10, 14, 18, \ldots$.

Now, Tarry's proof in 1900 that there was no 6×6 Eulerian square certainly added some much needed support to Euler's Conjecture, and the next half-century passed with no further progress made on the conjecture one way or the other, which if anything only served to validate Euler's intuition. But all that changed abruptly beginning in 1958 when Parker, Bose, and Shrikhande showed not only that Euler was wrong, but that he was about as wrong as he could possibly be. Beyond $n = 6$—that is, for $n = 10, 14, 18, 22, \ldots$—there *always* exists a pair of orthogonal $n \times n$ Latin squares for any n. Funnily enough, the first such pair that was found, by Bose and Shrikhande, was not a 10×10 square, but a 22×22 square! I urge you to read Martin Gardner's account of their remarkable discovery in [15].

The 10×10 Eulerian square that was found by Parker is shown in Figure 12.3 along with a black and white picture of a 10 foot $\times$ 10 foot Eulerian square that hangs in the Mathematics Department at Dartmouth College. You can see a color version of the Dartmouth Eulerian square at their departmental

00	47	18	76	29	93	85	34	61	52
86	11	57	28	70	39	94	45	02	63
95	80	22	67	38	71	49	56	13	04
59	96	81	33	07	48	72	60	24	15
73	69	90	82	44	17	58	01	35	26
68	74	09	91	83	55	27	12	46	30
37	08	75	19	92	84	66	23	50	41
14	25	36	40	51	62	03	77	88	99
21	32	43	54	65	06	10	89	97	78
42	53	64	05	16	20	31	98	79	87

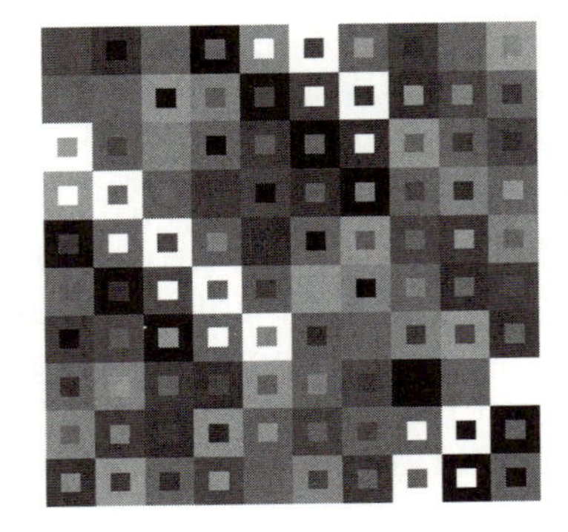

Figure 12.3 Parker's 10×10 Eulerian square and the Dartmouth College Eulerian square.

web site: www.math.dartmouth.edu/sphere. In the Dartmouth Eulerian or Graeco-Latin square, as well as in the painting on the cover of the November 1959 issue of *Scientific American* in which Martin Gardner announced the collapse of Euler's Conjecture, one of the two Latin squares is represented in 10 colors by the large outer squares, while the second Latin square is represented in the same 10 colors by the inner squares.

The almost 200-year-long story of Euler's famous conjecture was thus nearly at an end in 1958 with the dramatic discovery of Parker, Bose, and Shrikhande, but the story did have a denouement that was to last another 30 years due to the very close relationship between Eulerian squares and *finite projective planes.* It turns out—and it is not at all hard to show—that for a given value of n you can never have more than $n - 1$ mutually orthogonal $n \times n$ Latin squares. But to have as many as $n - 1$ of them is sometimes possible. For example, for $n = 5$, it is easy to construct 4 mutually orthogonal 5×5 Latin squares using the method described by Ball mentioned above, letting $a = 1$ for each of the four squares and successively letting $b = 1, 2, 3,$ and 4. On the other hand, for $n = 6$, since Tarry showed that even two mutually orthogonal Latin squares were impossible, certainly producing 5 of them is impossible. The relationship between Eulerian squares and finite projective planes, then, can be stated as follows: for a given value of n, there exists a finite projective plane of order n if and only if there are $n - 1$ mutually orthogonal $n \times n$ Latin squares. In other words, if you have a finite projective plane of order n, then you can construct $n - 1$ mutually orthogonal $n \times n$ Latin squares, and vice versa.

So, for example, we know there can't be a projective plane of order 6. However, for all other values of $n < 10$, there do exist the maximum number, $n - 1$, of $n \times n$ mutually orthogonal Latin squares; and, therefore, also, for all other values of $n < 10$, there do exist finite projective planes of order n. Thus, the case $n = 10$ became a critical one in the study of finite projective planes. Was there a finite projective plane of order 10, or not? When Parker found 2 orthogonal 10×10 Latin squares, that was certainly encouraging news, but still a long, long way from finding 9 mutually orthogonal 10×10 Latin squares.

Even to this day, no one has been able to find even 3 mutually orthogonal 10×10 Latin squares. But the story of Euler's Conjecture can perhaps be said to have finally and fully ended in 1988 when C. W. H. Lam proved that no projective plane of order 10 exists! This means, of course, that 9 mutually orthogonal 10×10 Latin squares don't exist either.

Let me end this chapter on Latin squares by showing you a puzzle I recently came across on the Internet. This one was provided by Eric Anderson, who produces a page of these puzzles daily at his web site: www.latinsquares.com.

Figure 12.4 A Latin square puzzle.

Problem 12.3 In Figure 12.4, a 6×6 chessboard has been divided into 6 pieces such that each piece contains exactly 6 squares of the chessboard. Also, 6 letters have already been placed on this chessboard. Complete a 6×6 Latin square by filling in the rest of the squares using the letters A, B, C, D, E, F so that *each* of the 6 pieces contains all 6 letters. There is a unique solution.

Solutions to Problems

Solution 12.1 This Latin square is shown in Figure 12.5.

7	2	5	0	3	6	1	4
6	1	4	7	2	5	0	3
5	0	3	6	1	4	7	2
4	7	2	5	0	3	6	1
3	6	1	4	7	2	5	0
2	5	0	3	6	1	4	7
1	4	7	2	5	0	3	6
0	3	6	1	4	7	2	5

Figure 12.5 An 8×8 Latin square generated using $a = 1$ and $b = 3$.

Solution 12.2 The revised Eulerian square is shown on the right in Figure 12.6. The magic sum for each row, column, and diagonal is 30. It should be clear why this works. Since in each row or column—and diagonal in this case—each Latin square contributes all four numbers 0, 1, 2, and 3, the sum will be $(0 + 1 + 2 + 3) + (0 \times 4 + 1 \times 4 + 2 \times 4 + 3 \times 4) = 6 + 24 = 30$. Note that the fact that in an Eulerian square all ordered pairs are distinct means that all sixteen numbers will be different in the revised square.

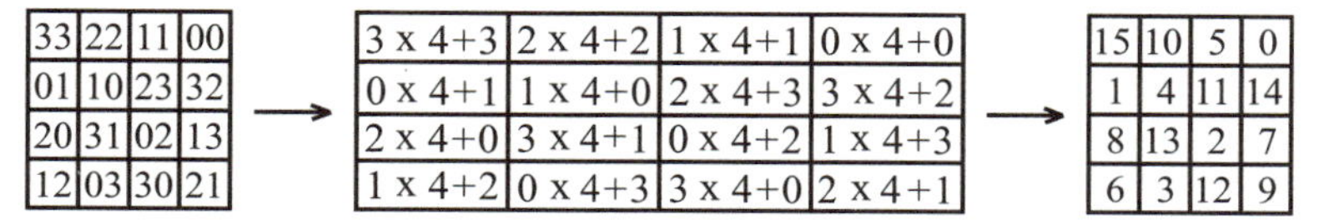

33	22	11	00
01	10	23	32
20	31	02	13
12	03	30	21

→

3 x 4+3	2 x 4+2	1 x 4+1	0 x 4+0
0 x 4+1	1 x 4+0	2 x 4+3	3 x 4+2
2 x 4+0	3 x 4+1	0 x 4+2	1 x 4+3
1 x 4+2	0 x 4+3	3 x 4+0	2 x 4+1

→

15	10	5	0
1	4	11	14
8	13	2	7
6	3	12	9

Figure 12.6 A magic Eulerian square.

Solution 12.3 The idea is to gradually work out what letter must be in each particular square. For example, in this puzzle, square $(6, 3)$ must contain an A since the other 5 letters are automatically excluded. Similarly, square $(5, 4)$ must contain an

F since there must be an F somewhere inside that piece, and an F can't go in the last column, so that's the only square available. This leaves only B and C to place in that particular piece. If we place C in square (6, 6), then the only square available in column 5 for C is square (5, 5), but then it is impossible to put a C anywhere inside the top piece on the left side, since the top two rows each already contain a C. Therefore, square (6, 6) must contain a B, square (6, 4) contains a C, and square (6, 1) contains a D. Now, the only place for a B in the top piece on the left side is in square (1, 5), and then the placement of the rest of the Bs follows easily. Continuing in this way yields the completed solution shown in Figure 12.7.

D	F	A	E	C	B
B	D	C	F	A	E
A	B	E	D	F	C
F	E	B	C	D	A
E	C	D	A	B	F
C	A	F	B	E	D

Figure 12.7 A Latin square such that each puzzle piece contains all six letters.

Polyominoes

DISSECTION PROBLEMS

There is a long history of geometric dissection problems in recreational mathematics and these problems often involve a chessboard in one way or another. Here are two problems that nicely illustrate the genre taken from a marvelous collection of puzzles, *Mathematical Puzzles of Sam Loyd* [12], America's foremost puzzle expert of the late nineteenth century. I urge you to look at the originals in the book. All of the charm of the original drawings and of Loyd's prose has been lost in the following versions, although the mathematics remains.

Problem 13.1 In this problem, a square tract of land containing four ancient oak trees, situated roughly as shown in Figure 13.1, is to be divided equally among four sons. Can this be done with absolute fairness by giving each son an identical piece of land, and one tree apiece?

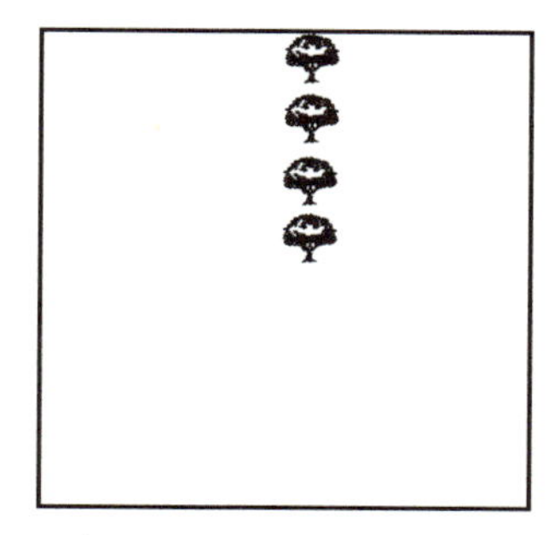

Figure 13.1 Divide the land into four identical pieces, each with an ancient oak.

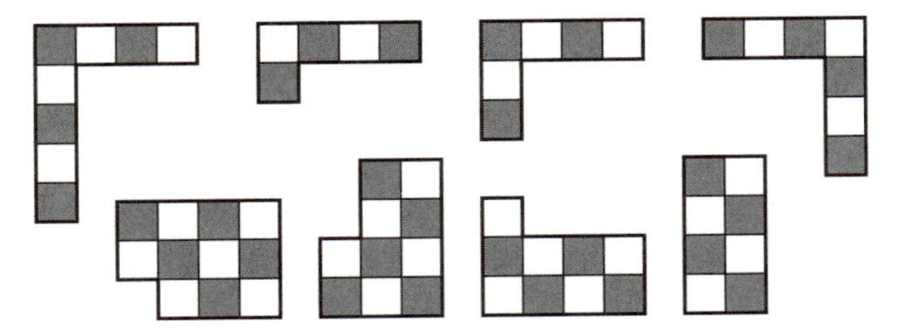

Figure 13.2 The eight pieces of a broken chessboard.

Sam Loyd is not the only author ever to imagine trees on a chessboard. Vladimir Nabokov, in *The Defense*, once wrote:

> He sat leaning on his cane and thinking that with a knight's move of this lime tree standing on a sunlit slope one could take that telegraph pole over there. . . .

Problem 13.2 In this second problem, Loyd tells an elaborate story of the young Dauphin of France breaking a chessboard into eight pieces over the head of the Duke of Burgundy. The eight pieces are shown in Figure 13.2, and the problem is to reconstruct the broken chessboard.

The Latin square dissection of the 6×6 chessboard in Problem 12.3 in Chapter 12 and Loyd's dissection of the 8×8 chessboard above seem quite similar. There is one interesting difference however. Loyd intentionally varied the size and shape of the pieces so that 'other ways' of doing the puzzle would be prevented; whereas, in the 6×6 Latin square puzzle it is of course necessary that all of the pieces have exactly 6 squares so that all of the letters A–F fit nicely into each piece. We will be particularly interested in dissections of chessboards such as this in which all of the pieces have the same size. We begin, therefore, at the beginning, with dominoes.

Dominoes

We have previously encountered the idea of covering chessboards with dominoes in Chapter 1 when we discussed the Domino Puzzle and Gomory's Theorem. It is obvious that the notions of covering a chessboard with dominoes, on the one hand, and dissecting a chessboard into 1×2 pieces, on the other hand, are completely equivalent notions. Here is a problem that introduces several very natural questions that we might ask about the ways in which chessboards can be covered using dominoes. These are questions that have been raised and investigated quite independently by a number of people, including, I should add, Kagen Schaefer, a student first mentioned in connection with Problem 2.6.

Problem 13.3 Cover a 5×6 chessboard with dominoes so that there are no *fault lines*—that is, no straight lines that are formed along the edges of dominoes that go all the way across the chessboard. This is the smallest rectangle for which this can be done, and it can be done in two different ways.

On the other hand, *any* covering of a 6×6 chessboard with dominoes will necessarily have a fault line. This is relatively easy to prove. Any horizontal or vertical line of a 6×6 chessboard splits the board into two pieces, each having an even number of squares. Therefore, any line of the chessboard that cuts a domino must in fact cut an even number of dominoes. And, of course, any domino can be cut by just one line, the line passing through its middle. Now, the 6×6 board has a total of 10 horizontal and vertical lines, 5 each way, and if *each* of these lines cuts a domino, then each of these 10 lines cuts at least 2 dominoes. But then, this would mean that there are at least 20 dominoes in all, and we know there are only 18 dominoes on this chessboard, so some line does not cut any domino. That is, there is a fault line.

This argument also works for the 4×4 chessboard. You might want to try it for the 8×8 board to see what happens. It is not at all hard to figure out exactly which rectangular chessboards necessarily have fault lines. The answer can be found in Martin Gardner's *New Mathematical Diversions* [15], and the counting argument used above for the 6×6 chessboard works perfectly for the non-trivial part of the proof given there.

There are several games that can be played on chessboards based on domino coverings. For example, two players can alternate placing a domino on the board, and the first player who cannot make a play loses the game. However, to be interesting this game should be played on an odd $\times$ odd chessboard, since the second player can always win any game played on an even $\times$ even board by always playing symmetrically to the first player; and on an odd $\times$ even board the first player can always win any game by first playing in the very center, and thereafter playing symmetrically to whatever play the second player makes. Gardner also suggests a variation that uses the notion of fault lines, and suggests two versions: in the *achievement* version, the winner of the game is the first player to complete a fault line; whereas, in the *avoidance* version, the loser is the first player who is forced to complete a fault line.

The Bricklayer's Problem

Dissecting rectangles into dominoes has a natural analogue in three dimensions. For example, could a bricklayer construct a $6 \times 6 \times 6$ cube entirely out of bricks of size $1 \times 2 \times 4$? This problem is not nearly as well known as the Domino Puzzle discussed in Chapter 1, but it deserves to be since the answer is *no*, it is not possible, even though 27 such bricks would seem to fit perfectly, since $27 \times 8 = 216 = 6^3$ after all—just as 31 dominoes would seem to fit perfectly to cover the 62 squares of the chessboard with 2 corners removed—and there doesn't seem to be any apparent reason why these bricks shouldn't go together is some way to make a cube. Moreover, another reason this problem should be more well known is that the proof of the statement about impossibility of the construction is the same as the proof of the Domino Puzzle: coloring!

Color each of the $1 \times 1 \times 1$ cells in the $6 \times 6 \times 6$ cube either black or white so that in the natural subdivision of the large cube into twenty-seven $2 \times 2 \times 2$ blocks, each of the eight $1 \times 1 \times 1$ cells in each block has the same color, and so that the colors of the $2 \times 2 \times 2$ blocks alternate between black and white in a standard three-dimensional checkerboard fashion. The first thing to note is that since there are 27 blocks, there are more blocks—and, hence, more $1 \times 1 \times 1$ cells—of one of the two colors than of the other color. The second thing to note is that no matter where a $1 \times 2 \times 4$ brick is placed in the cube, it will have 4 white $1 \times 1 \times 1$ cells, and 4 black $1 \times 1 \times 1$ cells. Therefore, it is impossible for the bricklayer to succeed.

Maybe all this problem needs to become more famous is simply a good name, which is why I have suggested calling it *The Bricklayer's Problem*. Martin Gardner addresses the question of exactly when a bricklayer would be well advised to avoid attempting to build an $n \times n \times n$ cube in [13].

Similarly, Jan Mycielski of the University of Colorado posed a problem to prove that for any $20 \times 20 \times 20$ cube constructed entirely out of $1 \times 2 \times 2$ bricks there must exist a straight line, perpendicular to one of the faces of the cube, which does not pierce any of the bricks [22]. While it is not geologically appropriate to call such a line a fault line, this is clearly a three-dimensional analog of the fault line domino problem for the 6×6 chessboard discussed above. A solution to Mycielski's problem appeared in [23].

De Bruijn's Theorem

It is clear that any rectangular chessboard with an even number of squares can be covered with dominoes. That is, for any $m \times n$ chessboard, as long as 2 divides either m or n, the board can be covered with 1×2 pieces. Can this be generalized? What about 1×3 pieces? 1×4 pieces? and so on? In other words, a natural general question to ask is: when can an $m \times n$ chessboard be covered with pieces of size $1 \times k$? Obviously, just as for dominoes, if k divides either m or n, then the board can be covered with $1 \times k$ pieces. Let's look at an example where k does *not* divide either m or n.

Let's look at the 6×6 chessboard and try to cover it with pieces of size 1×4. Here is an easy coloring proof modeled on the Bricklayer's Problem that shows this to be impossible. Color the squares of the 6×6 chessboard with 4 colors as shown in Figure 13.3. Note that with this arrangement of colors, no matter how you place a 1×4 piece on the board, it will contain one square of each color. But, there are 10 squares on the board

1	2	3	4	1	2
2	3	4	1	2	3
3	4	1	2	3	4
4	1	2	3	4	1
1	2	3	4	1	2
2	3	4	1	2	3

Figure 13.3 Golomb's coloring proof of de Bruijn's Theorem.

colored with color 2, and only 8 squares colored with color 4, so it is impossible to cover the 6×6 board with 1×4 pieces.

Solomon W. Golomb used this exact coloring argument to give a proof of *de Bruijn's Theorem*: the only time you can cover an $m \times n$ chessboard with pieces of size $1 \times k$ is in the simple situation when k divides either m or n [16]. The proof just amounts to showing in general that the periodic coloring used above with k colors always results in an unequal number of squares of some of the colors unless k divides m or n. In 1987, Stan Wagon, now of Macalester College in St Paul, Minnesota, wrote a remarkable paper that presented fourteen—yes, fourteen!—*different* proofs of de Bruijn's Theorem [30].

Wagon also raised an interesting question about whether de Bruijn's Theorem remains valid on a torus. Note that the coloring argument of Figure 13.3 doesn't work any longer on a torus, since a 1×4 piece placed on a toroidal 6×6 chessboard need not contain a square of each color. However, Golomb was able to come up with a new coloring that, for example, shows that even on a torus you can't cover an $m \times n$ chessboard with 1×4 pieces unless 4 divides either m or n—that is, de Bruijn's Theorem still holds, at least for 1×4 pieces. This new coloring for the torus is illustrated in Figure 13.4 for a 6×6 chessboard. In this coloring, each 1×4 piece on a torus always contains a square of a given color twice. But, there are 9 squares of each color, so it is impossible, even on a torus, to cover the 6×6 board with 1×4 pieces.

Anyone who has been keeping track of the number of times in this book that things which are periodic turn out to be interesting when put on a torus won't be at all surprised to learn

1	2	1	2	1	2
3	4	3	4	3	4
1	2	1	2	1	2
3	4	3	4	3	4
1	2	1	2	1	2
3	4	3	4	3	4

Figure 13.4 Golomb's proof of de Bruijn's Theorem on a torus for 1×4 pieces.

that de Bruijn's Theorem does fail on a torus, however. In other words, laying bricks on a torus really does give you a bit more creative flexibility. In the next problem, you are asked to find the smallest example of this phenomenon.

Problem 13.4 Show that it is possible to cover the 10×15 chessboard with 1×6 pieces on a torus, even though 6 divides neither 10 nor 15.

By the way, as de Bruijn himself originally proved, I should mention that de Bruijn's Theorem holds in all higher dimensions; and so, for example, an $a \times b \times c$ solid block can be constructed out of $1 \times 1 \times k$ bricks only when k divides at least one of a, b, or c.

Ross Honsberger discusses a very nice 'de Bruijn-like' construction problem using $1 \times 1 \times 3$ bricks in his book, *In Pólya's Footsteps* [19]. Obviously, you can't build a $7 \times 7 \times 7$ cube out of $1 \times 1 \times 3$ bricks—we don't need de Bruijn's Theorem to know this! However, you can build such a cube if you leave out a single $1 \times 1 \times 1$ cell from, say, an upper corner of the cube, because then you can trivially construct the bottom six layers by stacking $1 \times 1 \times 3$ bricks vertically, and it is then very easy to fill in the remaining 48 cells in the top layer with 16 bricks laid horizontally.

From where else in the original $7 \times 7 \times 7$ cube can you leave out a single $1 \times 1 \times 1$ cell, and still construct the remaining $7 \times 7 \times 7$ cube using $1 \times 1 \times 3$ bricks? A clever solution—using what should by now be a familiar-sounding approach—was provided by Will Self of Eastern Montana College.

The first thing to do is to label the cells of the cube, just as we do for chessboards, with ordered triples using $(0, 0, 0)$ for one corner—say, the front lower left corner. So $(6, 6, 0)$ would be the front upper right corner, $(6, 6, 6)$ would be the corner diametrically opposite $(0, 0, 0)$, and $(3, 3, 3)$ would be the cell in the very center of the cube. Note that for any $1 \times 1 \times 3$ brick, no matter how it is oriented in the cube, the sum of *each* of the three coordinates is divisible by 3. This is because two of the coordinates don't change within the brick, and the third set of coordinates are just three consecutive numbers. Here is an example of such a brick placed in the cube at cells $(2, 3, 5)$, $(2, 4, 5)$, and $(2, 5, 5)$; and the coordinate sums are, respectively, $2 + 2 + 2 = 6$, $3 + 4 + 5 = 12$, and $5 + 5 + 5 = 15$, all divisible by 3.

Now, the sum of each of the coordinates for *all* of the cells in the $7 \times 7 \times 7$ cube is also divisible by 3, since $0 + 1 + 2 + 3 + 4 + 5 + 6 = 21$. Therefore, if the $7 \times 7 \times 7$ cube has been constructed from $1 \times 1 \times 3$ bricks with one cell missing, the missing cell must have coordinates *each* of which is divisible by 3. That is, the only possible coordinates for the missing cell are 0, 3, and 6. This means the missing cell can only be in, say, the top layer at a corner, the middle of an edge, or in the center of the top layer; or it can be similarly placed on any of the six faces of the cube; otherwise—and this is the fact that makes Honsberger's problem so striking—the only place the missing cell can be is at $(3, 3, 3)$, the very center of the cube!

Trominoes

Returning to two dimensions and chessboards, let's do a variation of the Domino Puzzle where we use 1×3 pieces instead of 1×2 pieces. What else can these pieces be called except *trominoes*? Actually, we need to call them *straight trominoes* since there is a second kind of tromino, an 'L' tromino, as well. But, for the moment, we will consider only the straight kind. Remove a single square from an 8×8 chessboard; can we cover the remaining board with 21 straight trominoes?

To answer this question, we will use our standard labeling of the squares of our chessboard, where $(1, 1)$ is the lower left-hand corner square. Then, no matter where a 1×3 piece is placed, one of the coordinates remains constant within the tromino, and the other coordinate takes on three consecutive integers at the three squares; thus, the sum of the three numbers, for each coordinate, is divisible by 3. But, the sum of each coordinate for all of the squares of the 8×8 chessboard is also divisible by 3, since $1 + 2 + \cdots + 8 = 36$. Therefore, if the 8×8 board has been covered with 1×3 pieces with one square missing, the missing square must have coordinates each of which is divisible by 3. That is, the only possible coordinates for the missing square are 3 and 6. This means that the missing square can only be $(3, 3)$, $(3, 6)$, $(6, 3)$, or $(6, 6)$. I'll leave to you the relatively easy task of covering the chessboard with each of these squares removed in turn using only straight trominoes.

Golomb gave an alternative argument for this same result using the coloring shown in Figure 13.5—actually, he used red, white, and blue and called this a patriotic coloring [16]. Just as in Figure 13.3 for 1×4 pieces, each 1×3 piece in Figure 13.5 will always contain one square of each color. Therefore, since there are 22 squares of color 1, and 21 squares of each of the other two colors, the missing square must have color 1. Not only that, however, but also any square that is symmetric to the missing square must also have color 1. Now, if you look closely at Figure 13.5, you will see that the four shaded squares are the only squares of color 1 that are such that all of their symmetric partners are also colored with color 1. And, of course, these are the very same four squares found by our previous argument.

Here is a third, somewhat harder, but very pretty algebraic argument for the same result given by Joe Roberts in [26]. If we represent each square (i, j) on the chessboard by the algebraic expression $x^i y^j$, then the entire chessboard can be represented by the polynomial

$$f(x, y) = (1 + x + x^2 + \cdots + x^7)(1 + y + y^2 + \cdots + y^7).$$

Such a polynomial is called a *generating function* for the chessboard, since if we multiply the polynomial out completely each

1	2	3	1	2	3	1	2
3	1	2	3	1	2	3	1
2	3	1	2	3	1	2	3
1	2	3	1	2	3	1	2
3	1	2	3	1	2	3	1
2	3	1	2	3	1	2	3
1	2	3	1	2	3	1	2
3	1	2	3	1	2	3	1

Figure 13.5 Golomb's coloring solution for the Tromino Problem.

square of the board will be represented by exactly one term in the resulting expression. Note that in this set-up we are regarding the lower left-hand corner square as $(0,0)$ or x^0y^0, and, therefore, the upper right-hand corner square is $(7,7)$ or x^7y^7.

Now, a straight tromino placed on the chessboard is represented either by something like

$$x^2y^3 + x^3y^3 + x^4y^3 = x^2y^3(1 + x + x^2)$$

if the tromino is horizontal on the board, or by something like

$$x^6y^4 + x^6y^5 + x^6y^6 = x^6y^4(1 + y + y^2)$$

if the tromino is vertical on the board. Therefore, if the entire board has been covered with straight trominoes with one square missing, we can write

$$f(x,y) = (1 + x + x^2)g(x,y) + (1 + y + y^2)h(x,y) + x^ay^b,$$

where (a,b) is the location of the missing square.

Now, things get a bit more technical. In the complex numbers, 1 has three *cube roots*, namely 1, $-\frac{1}{2} + \frac{1}{2}\sqrt{3}\mathrm{i}$, and $-\frac{1}{2} - \frac{1}{2}\sqrt{3}\mathrm{i}$. We will write these three cube roots of 1 as 1, ω, and ω^2. You may want to check directly—using $\mathrm{i}^2 = -1$—not only that each of these numbers when cubed gives you 1 as advertised, but also that $(-\frac{1}{2} + \frac{1}{2}\sqrt{3}\mathrm{i})^2 = -\frac{1}{2} - \frac{1}{2}\sqrt{3}\mathrm{i}$, and that $1 + \omega + \omega^2 = 0$.

Therefore, plugging ω in for x and y in the polynomial $f(x,y)$, we get

$$(1 + \omega + \omega^2 + \cdots + \omega^7)^2 = \omega^{a+b}.$$

But

$$\begin{aligned}1 + \omega + \omega^2 + \cdots + \omega^7 \\ &= (1 + \omega + \omega^2) + \omega^3(1 + \omega + \omega^2) + \omega^6 + \omega^7 \\ &= \omega^6 + \omega^7 = 1 + \omega,\end{aligned}$$

so

$$\omega^{a+b} = (1 + \omega)^2 = 1 + 2\omega + \omega^2 = \omega.$$

Thus, $a + b \equiv 1 \bmod 3$.

Now, the clever part is to do this all over again, this time plugging in ω^2 for x and for y, which this time around gives us, after some work, $2a + b \equiv 0 \bmod 3$. Solving these two congruences for a and b yields $a \equiv -1 \bmod 3$ and $b \equiv 2 \bmod 3$. Thus, the missing square can only be $(2, 2)$, $(2, 5)$, $(5, 2)$, or $(5, 5)$. Since we labeled the lower left-hand corner square $(0, 0)$ in this particular set-up, these are the same four squares that are shaded in Figure 13.5.

Polyominoes

While it may have been most natural for us to think of covering chessboards with straight trominoes as an extension of the idea of a domino within the context of de Bruijn's Theorem, Solomon W. Golomb introduced the most general extension of the idea of a domino to the world in 1953 when he coined the term *polyomino* to mean any shape that consists of unit squares connected edge to edge—in other words, polyominoes are pieces of a chessboard! So, in the language of polyominoes, a single square would be called a *monomino*, two connected squares are the familiar *domino*, and since there are two ways to connect three squares, there is a *straight tromino* as well as the angled L *tromino*.

Golomb has a lovely inductive proof—starting with a 2×2 chessboard—that you can remove a single square from *anywhere* on an 8×8 chessboard, and then cover the remaining board with 21 L trominoes. So, the L Tromino Problem and the Straight Tromino Problem, which is the one we just finished

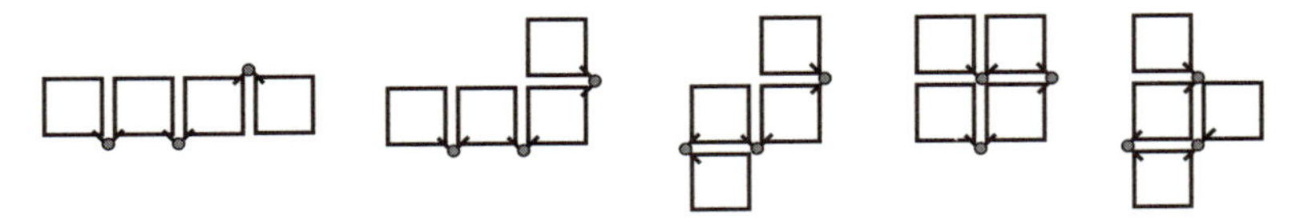

Figure 13.6 A set of hinged tetrominoes.

doing three different ways, have very different solutions. Polyominoes themselves appeared much earlier in the puzzles of Loyd and Dudeney, in Problem 13.2, for example, but there is no doubt that it was Golomb who made them what they are today. And you can do no better than to head straight to his book, *Polyominoes* [16].

There are five *tetrominoes*, and these were first drawn in Chapter 1 in Figure 1.11. Note that with tetrominoes we have two pieces that are actually different when they are turned over. So this becomes an important feature of polyominoes to keep in mind, not only can you move them and rotate them on a chessboard, you can flip them over as well. Here, in Figure 13.6, is another picture of the five tetrominoes showing a version of them that can be constructed in which the straight tetromino on the left has three *hinges* placed so that it can then be manipulated to form each of the other tetrominoes. Two students, Arden Rzewnicki and Sarah Eisen, and I wondered in general how many different ways you could hinge a straight polyomino [35].

It turns out that the number of different ways a straight polyomino having n squares can be connected with hinges is

$$2^{\lfloor (n-3)/2 \rfloor} + 2^{n-3}.$$

So, for example, with $n = 4$, there are $2^0 + 2^1 = 3$ different ways to connect the straight tetromino with hinges.

Rather remarkably, this same formula also gives us the number of different ways that the maximum number, $2n - 2$, of independent bishops can be placed on an $n \times n$ chessboard! So, for example, in Figure 10.14, the 14 independent bishops could be arranged in $2^2 + 2^5 = 36$ different ways on an 8×8 chessboard.

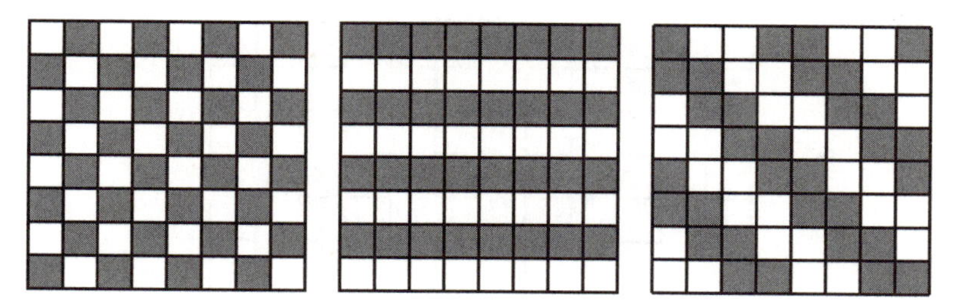

Figure 13.7 Colorings for three impossible tetromino coverings.

In Problem 1.4 in Chapter 1 we saw that the 8×8 chessboard could be covered by each of the tetrominoes, with the single exception of the Z tetromino. In a tour de force of coloring, Golomb demonstrates that, on the other hand, each of the following three coverings would be impossible using, in order, the three colorings of the 8×8 chessboard shown in Figure 13.7 [16].

It is impossible to cover an 8×8 chessboard with

1. fifteen T tetrominoes and one 2×2 tetromino;
2. fifteen L tetrominoes and one 2×2 tetromino;
3. one 2×2 tetromino and *any* combination of straight and Z tetrominoes.

Pentominoes

The most well known of the polyominoes—after dominoes, that is—are the *pentominoes*. There are 12 of them, and they are shown in Figure 13.8. Pentominoes are widely available commercially and come in a variety of sizes and materials. I highly recommend getting hold of a set, although cutting your own out of stiff paper works just fine. It is convenient and relatively unambiguous, though not terribly accurate, to refer to each pentomino by a letter of the alphabet; so the 12 pentominoes are: I, W, L, V, Z, P, N, T, Y, F, U, and X. I think you'll agree, N and F take some imagination, but you get used to them.

Figure 13.8 The 12 pentominoes.

Perhaps the best introduction to pentominoes for a beginner is to tackle the problem of showing that no matter where you place a 2×2 tetromino on the 8×8 chessboard, the rest of the board can then be covered with the 12 pentominoes. Since there are only ten fundamentally different positions in which the 2×2 tetromino could be placed, this amounts to doing ten different warm-up exercises. Golomb, however, has a clever way of reducing these ten cases to just three cases that can then be solved individually, a considerable saving in labor [16].

Since 12 pentominoes cover a total of 60 squares, there are several obvious rectangular chessboards for us to attempt to cover with a complete set of pentominoes.

Problem 13.5 Find coverings using the 12 pentominoes for each of the following chessboards: the 6×10, the 5×12, the 4×15, and the 3×20. In fact, you can solve the 6×10 and the 5×12 boards at one go by separating the 12 pentominoes into two groups of 6 pentominoes, and showing how to cover a 5×6 chessboard with each group of 6; and so, each group covers half of the larger chessboard. You can do a similar thing with the 4×15 chessboard, and have each group of 6 pentominoes cover congruent 'halves' of the board. As an additional challenge, you might also want to try to find solutions for each of these chessboards in which all 12 pentominoes touch the outer edge of the board.

Three Scaling Problems

A very well-known *scaling problem* involving pentominoes is the *Triplication Problem*: given a pentomino, use nine of the other pentominoes to construct a replica of the original pentomino three times as large. This can be done with each of the pentominoes. The idea is illustrated in Figure 13.9 for the X pentomino.

Figure 13.9 Triplication of the X pentomino.

The triplication problem is a very appealing one. There is something quite pleasant about seeing a familiar pentomino shape emerge from nine other pentominoes carefully fitted together. But not being able to use all 12 pentominoes is mildly unsatisfying. With that in mind, here is a version of this scaling problem I haven't seen before, although it seems quite natural. I call it the (4, 3)-*Replication Problem*, since the pentominoes are quite literally replicating themselves. The idea is to begin with a given pentomino, and then use the 12 pentominoes to construct a scale replica of the original pentomino that has been scaled—or stretched—by a factor of 4 in the horizontal direction and by a factor of 3 in the vertical direction. This is illustrated in Figure 13.10 for two different pentominoes, the X pentomino and the F pentomino. Note that the F pentomino actually creates two different replication problems because its original orientation matters; whereas, the orientation of the X pentomino doesn't make a difference in the outcome. The following problem asks you to do the (4, 3)-Replication Problem for the rest of the pentominoes. Note that the I pentomino has really already been done, since a horizontal $(4, 3)$-replication of I is the 3×20 rectangle, and a vertical $(4, 3)$-replication of I is the 15×4 rectangle.

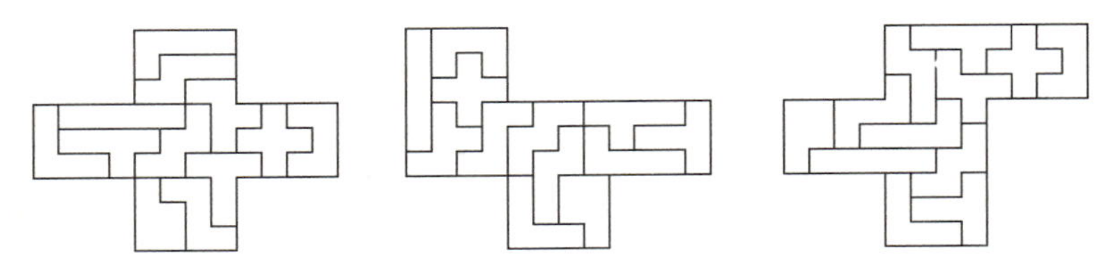

Figure 13.10 A $(4, 3)$-replication for the X pentomino and the F pentomino.

Problem 13.6 Solve the $(4, 3)$-Replication Problem for the V, W, L, N, P, T, U, Y, and Z pentominoes.

The motivation behind the $(4, 3)$-Replication Problem, of course, was to make the area of the scaled-up replica be exactly 12 times the area of the original pentomino. Another way to achieve much the same thing is to use the third dimension. However, this does require a set of pentominoes made out of $1 \times 1 \times 1$ cubes rather than flat two-dimensional pentominoes. Fortunately, such sets are also widely available.

The idea, then, for our third scaling problem is to begin with a given pentomino, and then use the 12 pentominoes to construct a three-dimensional scale replica of the original pentomino that has been scaled by a factor of 2 in each of the planar directions, and then by a factor of 3 in the vertical direction. Thus, the *volume* of the scaled-up replica will be exactly 12 times the volume of the original pentomino. I'll call this the *3D-Replication Problem*, although Golomb originally referred to it as one of a number of modeling problems [16]. This problem is illustrated in Figure 13.11 for the Z pentomino. The 3D-Replication Problem has a solution for all of the pentominoes, except for the W and X pentominoes; and the solution for the F pentomino is unique.

If you have a set of solid pentominoes—that is, a set made of $1 \times 1 \times 1$ cubes—then there are several obvious problems that

Figure 13.11 A 3D-replication of the Z pentomino.

are the three-dimensional analogues of the rectangle problems listed in Problem 13.5.

Problem 13.7 Using the 12 solid pentominoes, construct each of the following solids: the $3 \times 4 \times 5$ solid, the $2 \times 5 \times 6$ solid, and the $2 \times 3 \times 10$ solid.

Games

In Stanley Kubrick's movie *2001: A Space Odyssey*, the computer HAL plays a pentomino game on a 10×10 chessboard using a touch-sensitive computer screen. Sadly, these scenes ended on the cutting room floor and we now know HAL only as a chess-playing computer. The most obvious pentomino game, described by Martin Gardner [14], is played on an 8×8 chessboard, and two players take turns placing a pentomino on the board. The first player who is unable to play loses. The players can play from a common pile of pentominos, or divide the 12 pentominos in some fashion at the beginning of the game, or even be required to play a random pentomino drawn from a bag. I think an especially nice variation is to play this game on a torus! Another variation I like is to add a rule that the pentominoes cannot touch one another—or you might let them touch at a corner, but not along an edge—as in Figure 13.8. Since an 11×12 board is the smallest board on which you can place all 12 pentominoes with this no-touching rule, an 11×11 board is a good size for this particular variation.

Hexominoes and Beyond

The subject of polyominoes is a huge one and I have to stop somewhere. My intent in this chapter—as it has been throughout the book—has been only to whet your appetite. Just for fun, shown here in Figure 13.12 is a final chessboard, 51×58, that was covered by David Shields [16] with the 369 *octominoes*—that's all of them, and note that 6 of the octominoes have holes in the middle! What an amazing picture.

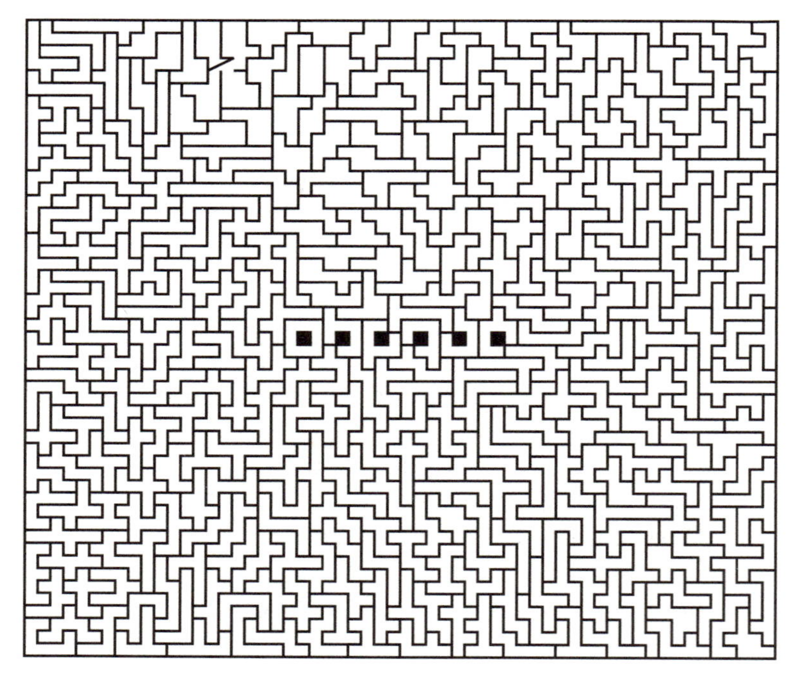

Figure 13.12 The 369 octominoes.

Solutions to Problems

Solution 13.1 Since each son must have one of the ancient oaks on his portion of the land, an acceptable solution is shown in Figure 13.13 with four identical pieces of land. Not very practical, but it is a solution.

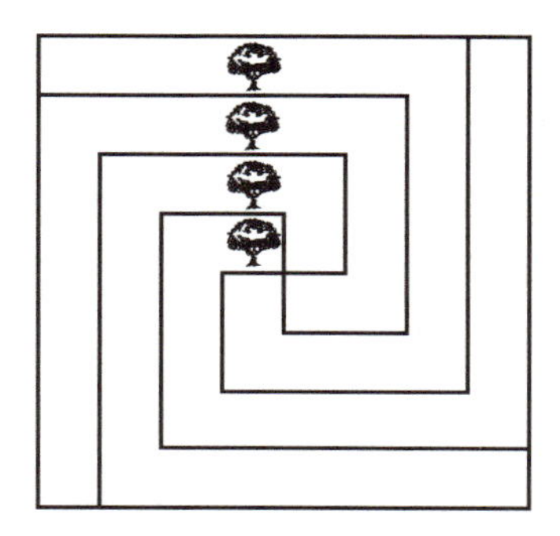

Figure 13.13 A perfectly fair dissection of the property.

Solution 13.2 The reconstructed chessboard is shown in Figure 13.14.

Figure 13.14 The reconstructed chessboard.

Solution 13.3 The two solutions are shown in Figure 13.15.

Solution 13.4 Here in Figure 13.16 is Golomb's example. Note that the pattern repeats in a knight's $(3, 2)$-move.

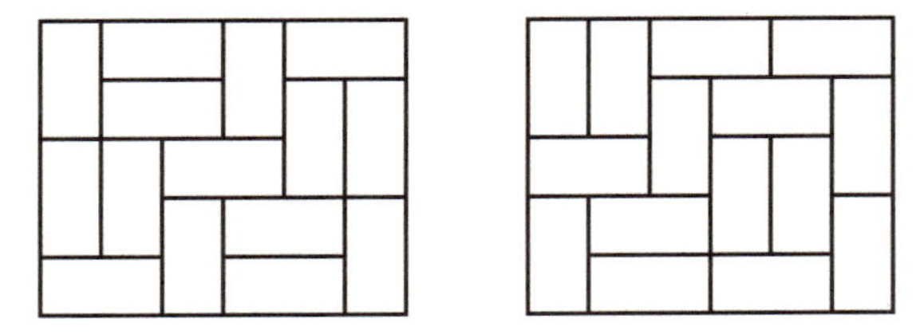

Figure 13.15 Two fault-free domino coverings.

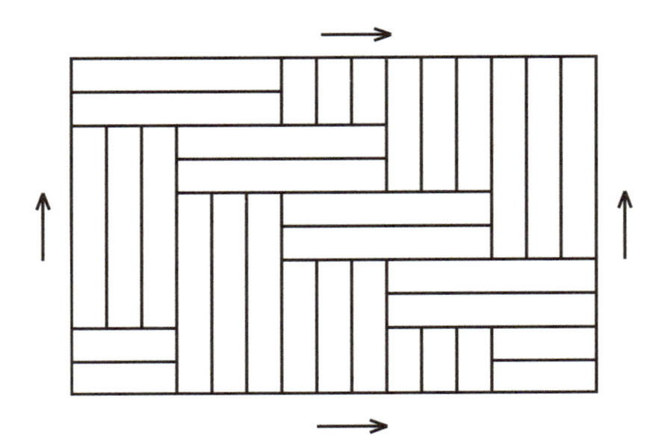

Figure 13.16 Golomb's counterexample for de Bruijn's Theorem on a torus.

Solution 13.5 Two solutions for the 6×10 chessboard are shown in Figure 13.17. The first solution is just one I constructed having decided to place the X near the center. The second, however, is one of only two solutions for the 6×10 chessboard that have all 12 pentominoes touching the outer edge. This second solution also has the property that the left half is congruent to the right half. Incidentally, the 6×10 board can also always be solved—if you want to make things harder—by tossing the 12 pentominoes on a table randomly, and then not allowing yourself to turn any of the pieces over.

Two solutions for the 5×12 board are shown in Figure 13.18. The first has the property that the left half is congruent to the right half; whereas, in the second solution all 12 pentominoes touch the outer edge.

Alternatively, the 6×10 chessboard and the 5×12 chessboard can each be solved by doing two 5×6 boards with one

Figure 13.17 Two solutions for the 6×10 chessboard.

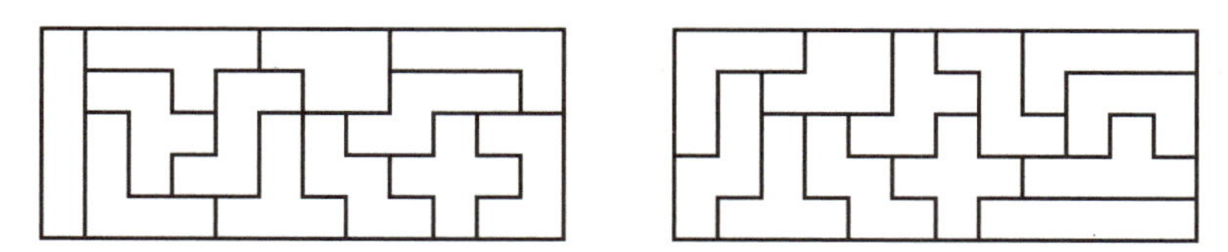

Figure 13.18 Two solutions for the 5×12 chessboard.

set of 12 pentominoes, and then combining the two half-boards appropriately. This solution is shown in Figure 13.19. You simply bring these two boards together for a 5×12 solution, and you rotate them first for a 6×10 solution.

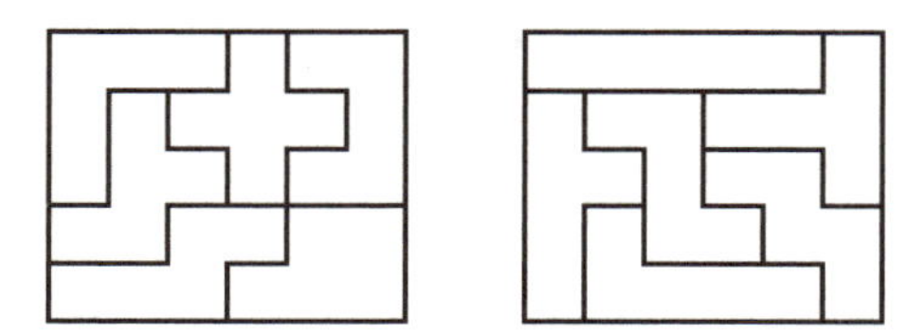

Figure 13.19 A solution for two 5×6 chessboards with one set of 12 pentominoes.

A solution for the 4×15 board is shown in Figure 13.20. This solution has two congruent halves, and all 12 pentominoes touch the outer edge.

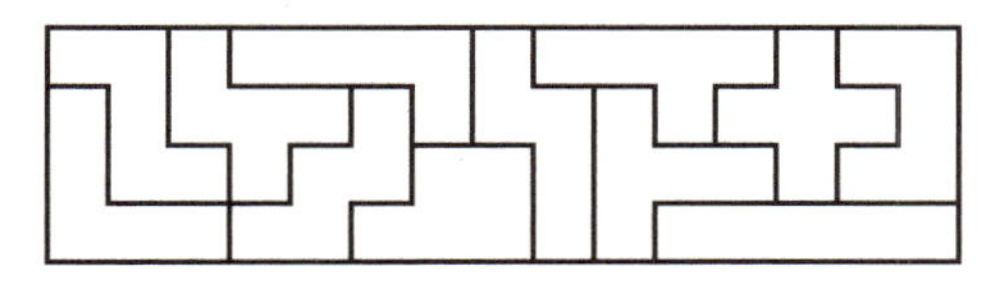

Figure 13.20 A solution for the 4×15 chessboard.

Finally, there are only two possible solutions for the 3×20 chessboard, and these are both shown in Figure 13.21. That there are only two solutions can be proved by first showing that it is impossible to split the 3×20 board into two smaller rectangular boards and cover both boards with 12 pentominoes. This allows you to prove that V has to be at one end of the 3×20 board, and U together with X at the other end, and then the proof is reduced to considering a manageable number of cases. Needless to say, all 12 pentominoes touch the outer edge in each solution. Note that these two solutions are essentially the same, and differ only in that the shaded portion, being symmetric, has been rotated.

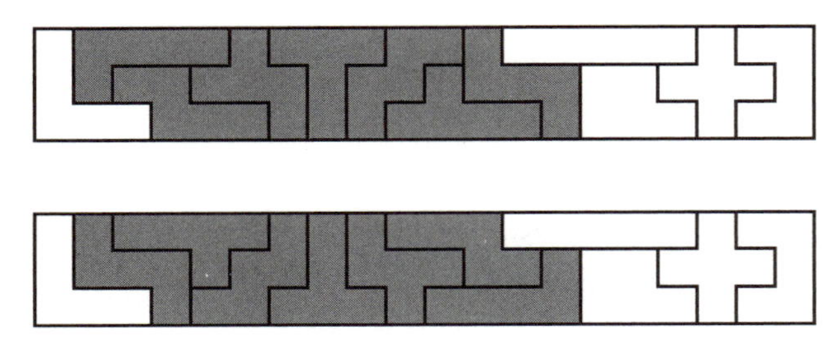

Figure 13.21 The only two solutions for the 3×20 chessboard.

On the other hand, if you place the 3×20 chessboard on a cylinder, then the U and X no longer need to be at one end. Here, in Figure 13.22, is a different pentomino covering of the 3×20 chessboard on a cylinder.

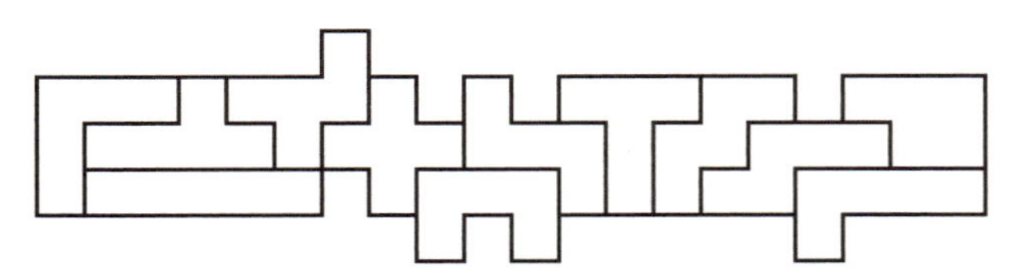

Figure 13.22 A solution for the 3×20 chessboard on a cylinder.

Similarly, you can also do two 3×10 chessboards on a cylinder, as shown in Figure 13.23, with a single set of 12 pentominoes.

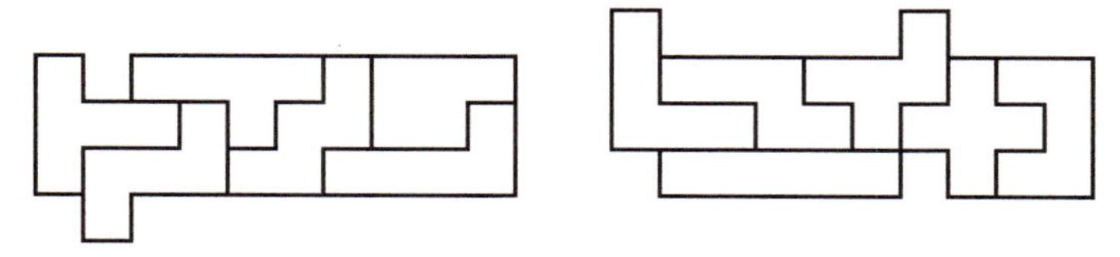

Figure 13.23 A solution for two 3×10 chessboards on a cylinder.

Solution 13.6 Solutions are shown in Figure 13.24. Note that V and W have the same shape in either orientation, and so only require one solution. All of the others require two.

Figure 13.24 (4,3)-replications for V, W, L, N, P, T, U, Y, and Z.

Solution 13.7 Solutions for the $3 \times 4 \times 5$ solid and the $2 \times 3 \times 10$ solid are shown in Figure 13.25. Note that in the solution for the $3 \times 4 \times 5$ solid, the U pentomino is in the center, and hence is not visible in the diagram, but its position is nonetheless easy to determine. The $2 \times 5 \times 6$ solid can be constructed in two separate layers using Figure 13.19.

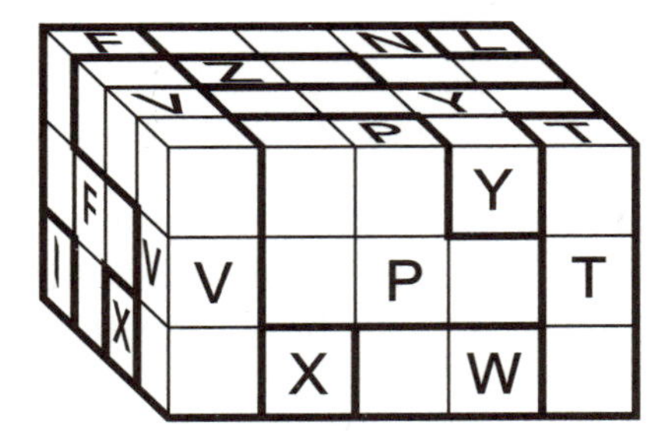

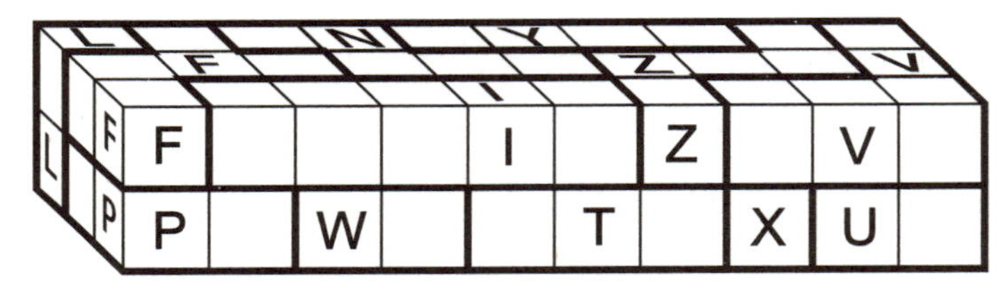

Figure 13.25 Solutions for the $3 \times 4 \times 5$ solid and the $2 \times 3 \times 10$ solid.

References

[1] Ahrens, W. 1901 *Mathematische Unterhaltungen und Spiele.* B. G. Teubner, Leipzig.

[2] Ball, W. W. R. 1987 *Mathematical Recreations and Essays*, 13th revised edn. Dover, New York.

[3] Balof, B. & Watkins, J. J. 1996 Knight's tours and magic squares. *Congressus Numerantium* **120**, 23–32.

[4] Burger, A. P., Mynhardt, C. M. & Cockayne, E. J. 1994 Domination numbers for the queen's graph. *Bulletin of the ICA* **10**, 72–82.

[5] Campbell, P. J. 1997 Gauss and the eight queens problem: a study in miniature of the propagation of historical error. *Historia Mathematica* **4**, 397–404.

[6] Cockayne, E. J. 1990 Chessboard domination problems. *Discrete Mathematics* **86**, 13–20.

[7] Cockayne, E. J., Favaron, O., Payan, C. & Thomason, A. 1981 Contributions to the theory of domination, independence, and irredundance in graphs. *Discrete Mathematics* **33**, 249–258.

[8] Cockayne, E. J. & Hedetniemi, S. T. 1986 On the diagonal queens domination problem. *Journal of Combinatorial Theory* A **42**, 137–139.

[9] Demirörs, O., Rafraf, N. & Tanik, M. 1992 Obtaining n-queens solutions from magic squares and constructing magic squares from n-queens solutions. *Journal of Recreational Mathematics* **24**(4), 272–280.

[10] Dudeney, H. E. 1958 *Amusements in Mathematics.* Dover, New York.

[11] Fricke, G. H., Hedetniemi, S. M., Hedetniemi, S. T., McRae, A. A., Wallis, C. K., Jacobson, M. S., Martin, H. W. & Weakley, W. D. 1995 Combinatorial problems on chessboards: a brief survey. *Graph Theory, Combinatorics, and Applications* **1**, 507–528.

[12] Gardner, M. 1959 *Mathematical Puzzles of Sam Loyd.* Dover, New York.

[13] Gardner, M. 1976 *Scientific American*, vol. 2 (February).

[14] Gardner, M. 1990 *Mathematical Magic Show,* revised edn. Mathematical Association of America, Washington, DC.

[15] Gardner, M. 1995 *New Mathematical Diversions.* The Mathematical Association of America, Washington, DC.

[16] Golomb, S. W. 1994 *Polyominoes,* 2nd edn. Princeton University Press, Princeton.

[17] Hare, E. O. & Hedetniemi, S. T. 1987 A linear algorithm for computing the knight's domination number of a $k \times n$ chessboard. *Congressus Numerantium* **59**(5), 115–130.

[18] Haynes, T. W., Hedetniemi, S. T. & Slater, P. J. (eds) 1998 *Domination in Graphs.* Marcel Dekker, New York.

[19] Honsberger, R. 1997 *In Pólya's Footsteps.* The Mathematical Association of America, Washington, DC.

[20] Kraitchik, M. 1953 *Mathematical Recreations.* Dover, New York.

[21] Lemaire, B. 1973 Covering the 11×11 chessboard with knights. *Journal of Recreational Mathematics* **6**(4), 292.

[22] Mycielski, J. 1970 *The American Mathematical Monthly* **77**, June/July, p. 656.

[23] Mycielski, J. 1971 *The American Mathematical Monthly* **78**, August/September, p. 801.

[24] Petković, M. 1997 *Mathematics and Chess.* Dover, Mineola, New York.

[25] Pickover, C. A. 2002 *The Zen of Magic Squares, Circles, and Stars.* Princeton University Press, Princeton and Oxford.

[26] Roberts, J. 1992 *Lure of the Integers.* The Mathematical Association of America, Washington, DC.

[27] Schwenk, A. J. 1991 Which rectangular chessboards have a knight's tour? *Mathematics Magazine* **64**, 325–332.

[28] Stewart, I. 1992 *Another Fine Math You've Got Me Into.* Freeman, San Francisco, CA.

[29] Stewart, I. 1971 Solid knight's tours. *Journal of Recreational Mathematics* **4**(1), 1.

[30] Wagon, S. 1987 Fourteen proofs of a result about tiling a rectangle. *The American Mathematical Monthly* **94**, 601–617.

[31] Watkins, J. J. 1997 Knight's tours on triangular honeycombs. *Congressus Nunerantium* **124**, 81–87.

[32] Watkins, J. J. 2000 Knight's tours on cylinders and other surfaces. *Congressus Numerantium* **143**, 117–127.

[33] Watkins, J. J. & Hoenigman, R. 1997 Knight's tours on a torus. *Mathematics Magazine* **70**(3), 175–184.

[34] Watkins, J. J., Ricci, C. & McVeigh, B. 2002 King's domination and independence: a tale of two chessboards. *Congressus Numerantium* **158**, 59–66.

[35] Watkins, J. J., Rzewnicki, A. & Eisen, S. 1999–2000 Hinged polyominoes, caterpillars, and bishops. *Journal of Recreational Mathematics* **30**(4), 239–245.

[36] Weakley, W. D. 1995 Domination in the queen's graph. *Graph Theory, Combinatorics, and Applications* **2**, 1223–1232.

[37] West, D. B. 2000 *Introduction to Graph Theory*, 2nd edn. Prentice Hall, Upper Saddle River, NJ.

[38] Wilson, R. J. 1996 *Introduction to Graph Theory*, 4th edn. Addison-Wesley, Boston, MA.

[39] Wilson, R. J. & Watkins, J. J. 1990 *Graphs: An Introductory Approach*. Wiley, New York.

[40] Yaglom, A. M. & Yaglom, I. M. 1964 *Challenging Mathematical Problems With Elementary Solutions*, Volume 1: *Combinatorial Analysis and Probability Theory*. Holden-Day, San Francisco, CA.

[41] Zaslavsky, C. 1999 *Africa Counts: Number and Pattern in African Culture*, 3rd edn. Lawrence Hill Books, Chicago, IL.

Index